GOD AND THE ATOM

GOD AND THE ATOM

RONALD KNOX

CLUNY
Providence, Rhode Island

CLUNY EDITION, 2024

For more information regarding this title
or any other Cluny Media publication,
please write to info@clunymedia.com, or to
Cluny Media, P.O. Box 1664, Providence, RI 02901

VISIT US ONLINE AT WWW.CLUNYMEDIA.COM

IMPRIMATUR: E. Morrogh Bernard, *Vic. Gen.*
WESTMONASTERII, DIE 27 OCTOBRIS 1945

This book is produced in complete conformity with the war economy agreement.
(NOTE FROM THE 1945 SHEED & WARD EDITION)

Cover design by Clarke & Clarke
Cover image: John Constable, *Study of Clouds* (detail),
1822, oil on paper, laid on canvas
Courtesy of Wikimedia Commons

CONTENTS

"To let a creed, built in the heart of things.
Dissolve before a twinkling atomy!"

—WORDSWORTH, *Excursion*

To Hubert Van Zeller

Monk of the Order of St. Benedict

Author's Note

Any accuracy discoverable about the scientific allusions
in this book must be attributed to the kind offices of
Sir Edmund Whittaker, Professor of Mathematics in
the University of Edinburgh, who was at pains
to look through the chapters in question.

TRAUMA

I

Hiroshima

AT A moment when it seemed as if all our capacity for surprise were already exhausted, one day last August, we opened the paper to find that we were wrong. Something had happened compared with which the General Election, and even Victory Day, would probably seem unimportant in perspective. A Japanese town, rather more populous than Southampton, had suddenly ceased to exist.

We knew that scientists had been busy with the atom; that it had lost the privilege of infinitesimality, and yielded, like everything else, to analysis. But this was only theoretical knowledge, surely? There were people, indeed, who talked of "bombarding" the atom, and thereby splitting it; who hinted darkly that this process might mean the releasing of a new source of power, destined, perhaps, to throw all others into the shade. But of course it would come slowly, like television; at present, all this talk of atoms was merely one more way of impressing and confusing the lay mind, like Einstein's universe. And all the time, behind our backs, men of science were working feverishly at unmentionable researches, machinery was being moulded by patient manufacturers who did not know what it was for, millions of money were being lavished on a weapon that might never come into use, only came into use when it was nearly too late. With something of that suddenness which was so awfully characteristic of the event it chronicled, the news burst upon our breakfast-tables and left us numb.

Our first reaction, probably, was one of fear. True, the secret had been well kept; it was in the right hands. But secrets can be betrayed, or guessed, and it is not often, or for long, that right hands monopolize the power to destroy. An indefinite period of universal peace, our confident assumption in 1918, is only a frantic aspiration in 1945; we pictured the next war, fought with such weapons as would make the rocket-bomb seem what in truth it was, last year's model. Even if there were no more wars, it was too much to hope there would not be rioting and civil disturbances; what hideous weapons, in a few years' time, might be within reach of the gunman and the hooligan! Was our civilization, after all, to turn Troglodyte?

That first reaction was followed, in many minds, by a sense of shame. True, it had all happened very far away, in a part of the world where devastating earthquakes are a common experience, evoking no rumble here, and hardly a gesture of commiseration. True, it had happened (with a certain poetic justice) to the country which bombed Pearl Harbour without an ultimatum. True, *we* were not consulted; even our Government, it seemed, was only an accessory before the fact. Yet a very general bewilderment was registered, a moral bewilderment; would it not have been possible to introduce the new weapon by dropping it on some unfrequented mountainside, and asking the Japanese whether they wanted to go on with the war after that? All the scruples some people had felt about the bombing of large target areas in Germany began to revive. We had told ourselves that *this* incident was, after all, only the logical development of *that* incident; it was difficult to know where you ought to draw the line. The thin end of the wedge had passed the scrutiny of the European conscience; and now, with appalling suddenness, the wedge was driven home; sixty thousand people wiped out, with no moment allotted to them to breathe a prayer!

I remember hearing of a hospital nurse who exclaimed, on hearing of the accident to R.101, "Now I know there is no God." It is not given to all of us to have our theological convictions so nicely balanced on the razor's edge. But we walk, even the most favoured of us, by faith and not by sight; the realities of the supernatural world do not produce the same clear impact upon us as the evidence of the senses. And many of us, to whom our religion means—or so we hope—more than anything else in life, are nevertheless liable to the chill whispers of doubt; they strike at the roots of us, now and again, like a draught catching an unsound tooth. To some people, I think, the third effect of Hiroshima will have been a glimpse, unreasonable but not unaccountable, down that dark vista which opens before the mind if it thinks of a world without God.

Not in the crude sense intended by the hospital nurse, that if there were a God he would not allow such things to happen. He allows earthquakes to happen, not less destructive, almost equally sudden; with nothing apparently done to deserve them, with no military advantage to apologize for them. No, the subtle shock given to our convictions arose from a different consideration—was Man, after all, the master of his own destinies more than we had supposed? The sense of our own inadequacy is a powerful suasion which tends to drive us back upon God. All those terrific exposures of human impotence and human ignorance which you read in the final chapters of Job are still valid after all these centuries of progress; how tiny is our contribution, after all, to the government of the universe; how inadequate, even, our comprehension of the principles on which it is governed! Our own littleness makes us acquiesce without protest in the doctrine which tells us that a Mind greater than ours, a Strength greater than our own, has been, and is, at work. Across that bright sun of our believing a shadow fell; the shadow of an American airman, lackadaisically humming a tune, who carried such a freight as would make life impossible over four square miles of the earth's surface.

He was the symbol of a catastrophic leap in the history of human achievement. Our efforts to gain the mastery over brute nature had been belittled for us hitherto by the slow tempo of their emergence. Newton's apple might be seen, afterwards, as the beginning of great things; at the time, it seemed no more than a comic accident. When James Watt's kettle boiled over there were no paragraphs in the papers. Every fresh stage of scientific advance seemed conditioned by a leisurely process of trial and error; we could look back, with a smile, on the struggles of the early motorist. Now for the first time (or so it seemed) Athene had sprung ready-armed from the brain of Zeus; what would have read, yesterday, like an extract from one of Mr. Wells's early romances had become, today, an accomplished fact. The effect, to be sure, was largely created by illusion; the process of trial and error had been there, but we had had no news of it, because the Allied Governments had very prudently determined that no news should be available. And, after all, the same American airman had attained a very respectable record of destructiveness without these new-fangled methods, yesterday and the day before.... No, but the point was the *scale* of the thing. We shall know, before many years have passed, how readily accessible the new force is, in what quantities it can be employed, to what purposes it can be harnessed. But meanwhile one thing is evident about it; it is a jump, not a mere step, for better or worse or both, in the history of civilization.

Moreover, it is difficult not to feel a kind of fabulous quality about this business of splitting the atom. An earlier age had made so certain that there was a minimum of divisibility beyond which you could not go; the molecule, yes, it might be possible, at least in thought, to resolve the molecule into its component parts, but beyond that lay something smaller yet, completely indivisible; it was christened accordingly "the atom"—the unsplittable thing. So we elders were taught at school; and now it seems we were only listening to conjurer's patter, leading up to the great illusion! But the mere trick of language still has its effect on us; to split the atom still *sounds* like a miracle. If we lived in an age when diabolical magic was generally believed in and freely traced, we should find ourselves wondering what the fiend would do now, to assert his superior ingenuity. Unaccustomed to such ways of thought, we are assailed at the back of our minds by a more intimate scruple. Is not Man, when he suddenly thus rises above his own level, infringing the prerogative of God himself?

Illusion? Of course it is an illusion. The God of religion (as distinct from the "God" of certain modern philosophies) is essentially the Creator of the visible world, and we men claim no power to do more than rearrange the material he has put at our disposal. The inventor is, by definition, only the finder; he presupposes the Hider. The man who discovered the latest possibilities of pitchblende is in no better position to assert or to deny the existence of a Deity than the man who first coaxed fire out of a flint. That is evident to all of us, on the level of consciousness. But at that lower level where the imagination absorbs impressions dimly received, feeds on the swill of our reflective thought, there may be a tiny shifting of balance; we felt certain, till now, we could not do without God as the explanation of things, now we are not quite so certain. Brute matter could not enter the lists as a rival explanation; a negative thing, mere potentiality. But this Force lurking at the very root of matter, that could so terrifically signalize its presence after all those centuries of patient imprisonment—that was quite a different affair. It stirred restless currents in the depths under the surface of the mind; it called to that instinct of idolatry which still lies hidden in the most sophisticated of us. From the old Roman augurs down to Henri Bergson, we have had the temptation to worship the *numen*, the Life-Force at the back of things. And Hiroshima was its epiphany.

So I, at least, read the lesson of that August morning. It is obviously open to anybody to say that all this is fantastic nonsense; you do believe in God or you don't, and the un-lamented disappearance of sixty thousand Japanese citizens will make you neither more nor less Christian. But am I quite wrong in suspecting that we are all, to some extent, psychologically conditioned—I do not say in our beliefs, for it might get me into trouble with the theologians, but in our proneness to believe? And if this kind of psychological conditioning is a real fact, is there not reason to fear that a *trauma* (if that is the word the modern psychologists would have me use) may be set up by the rude impact of a decisive event, at once scientific and historical, like the coming of the atom-bomb? That the atheists of to-morrow may be in the making, and all unconscious of it, as they read through two dozen lines of cold print about something that has happened two continents away? And all the more so because, as I have been trying to indicate, this new departure hits us in three separate ways, catches us out just where our imperfect faith is most open to attack? It strikes at our sense of cosmic discipline; it strikes at our optimism; it strikes at our confidence in the validity of our own moral judgements.

Christians adhere to God by faith, hope and charity. That is, they believe in the existence of a supernatural world; they conceive of the supernatural world as having its own purposes which are being worked out, obscurely to us, on this makeshift scene of time; they believe in a code of right and wrong, to be lived by here and now, which has supernatural sanctions, and eternal consequences. Or, if you will, Christians believe in a God of whom St. Paul said, "Of him, and through him, and to him are all things." His existence gives our existence its explanation, tells us how we got here; its meaning, tells us why things work out just so; its purpose, tells us what we are here for. Notoriously, that explanation is given only on general lines; believe in God, and existence is still a riddle, faith is still needed if we are to hold that all is for the best, even the life of moral action is clouded, not seldom, by scruples over an apparent conflict of duties. But in the main we can be content; and those to whom the sense of God's existence is most vivid, holy people, seem to achieve both a lightness of heart and a lightness of touch which indicates that their lives are integrated—they are no longer worrying about the ultimate things.

The atom, which now stands in the forefront of the news, is a symbol of our misgivings under all those three heads. All our classical notions of proving the existence of God are expressed in terms which imply that strict uniformity, strict causality, are to be found in nature; and the atom (so we are assured by the most intimate observers of its habits) seems to behave in a way that is literally unpredictable; it is not merely that we, with our limited knowledge, cannot foresee the moment at which a tiny explosion will take place in the minute underworld of creation, but that moment is literally undetermined, subject to no law but that of chance. Again, optimism does not feed on air. In theory, a blind act of faith should be enough to assure us that all is for the best, in a world Providentially governed. But in fact, almost everybody has his own philosophy of history, and likes to trace in the world-events of the past the influence of a Power making for righteousness. In particular, the men of our own age have been obsessed with the idea of a world in which freedom and common humanity were becoming, from century to century, more firmly established. But the advent of a new weapon, destructive on a scale hitherto unknown, seems to alter the whole perspective of historical probabilities; men who till yesterday were boldly prophesying a golden age are now wistfully hoping for it. Again, a man braces himself more readily for the moral struggle if he believes that right and wrong are easily distinguishable, and that the society to which he belongs adheres, at least in its major decisions, to the cause of right. Rouse in him the very suspicion that his country, when it goes to war, makes use of any and every expedient to achieve its end, and you tarnish the niceness of his own conscience; he is more ready to shrug-his shoulders and say, "After all, one must live." The shadow of that American airman strikes him with a faint, simultaneous chill of doubt, of despair, and of guilt.

God and the Atom

One of our cheap newspapers, during the week after the bombing of Hiroshima, had the inspiration of dating its issues by the fifth or sixth day, or whatever it might be, "of the Atomic Age." I hope it is not a sign of insecure mental balance to be disturbed by such puerilities. The cheap newspapers, there is every reason to fear, both reflect and form the mental outlook of their readers. This kind of suggestion—that we belong to a new period, that a line has been drawn across the page of history—insinuates itself very easily into the mind, and with formidable effects. Not for nothing did the statesmen of the French Revolution set about revising the calendar; these tiny details in the build-up of our world-outlook have power to prejudice the mind. If we are going to think of our age as wholly discontinuous with the past, and to boast of it, we shall not even have the gloomy satisfaction of profiting by experience. No bad thing, perhaps, that the world should feel, at the end of a long war, as if it had been through a baptism of blood, and emerged regenerate. But the clean sheet must be a white sheet of penitence; we must acknowledge responsibility for the past, even if we are inclined to blame our elders for it; *peccavimus cum patribus nostris*. The affectation of being a different kind of animal can have no other result than to leave us at the mercy of rash counsellors. Heaven preserve us all from a baptism of uranium.

Whatever the uses, good or evil, for which we shall

see the atom employed in the next few years, I have little doubt that it will be launched as a fresh bombshell against the structure of religious orthodoxy. In the first instance, by the enemies of religion; for religion, let the area of its influence shrink as it may, still has its enemies, determined to clean up the last pockets of resistance. The current of world-affairs in eastern Europe is setting ominously against us; bestowing a perfunctory handshake of recognition upon any ecclesiastical body which offers to toe the line, State-endowed atheism has been making strides everywhere, and England is coming in for the late effects of it, as she came in for the late effects of Jacobinism in France. And I suspect that the atom will be the totem of irreligion tomorrow, as the amoeba was yesterday. Meanwhile, we have to reckon not only with the attacks of our enemies, but with the inadequate apologias of faint-hearted friends. There will be an intensified demand for the kind of apologetic which gives up the notion of religious certainty, and attempts to rally the sporting spirit of our compatriots in favour of a balance of probabilities. There will be fresh attempts to dissociate natural theology altogether from our experience of the world around us, to concentrate more and more on arguments derived from the exigencies and the instincts of human nature itself. Meanwhile, the seminary-trained theologian, with all the wisdom of the centuries at his finger-tips, will find himself more than ever talking a strange language, more than ever at cross-purposes with the shibboleths of an Atomic Age. So it will go on, I suppose, till we find someone with enough courage, enough learning, enough public standing to undertake the synthesis; there is a battle royal, long overdue, which has still to be fought out at the level of academic debate.

To such destinies I do not aspire; age has brought me distaste, without bringing me competence, for the arena. The scope of this essay is strictly limited to the situation I have been trying to outline thus far. It does not hope to meet the arguments which would be put forward in any reasoned statement of the case against Theism, or the case against Thomism. It is an attempt to dispel an atmosphere unfriendly to the appeal of religion, an atmosphere psychologically conditioned by the prominence which will necessarily be given to atomic power in the popular literature of the coming decade. Say, if you will, that it is an attempt to analyse away the Hiroshima-complex in the minds of well-disposed but muddle-headed people; and perhaps, in a sense, having analysed, to sublimate it. In more homely terms, it is an attempt to pull the skeleton out of the cupboard and determine whether it is not, after all, a turnip-ghost.

Against one misinterpretation I am anxious to guard myself. It may easily be suspected that I am seizing upon a moment when Science appears as the destroyer, not as the benefactor of mankind, to cast discredit upon the whole scientific outlook. Against that suspicion would protest, and with energy. In this century it is the camp-followers of science, not its votaries, that come forward as the critics of religion; and an attack, from our side, against the scientific fraternity as such would be as misplaced as it would be short-sighted. It is a mere accident of history that atomic force has been born into the world with an ill name; as much an accident of history as if the first motor had been a tank, or the first aeroplane a bomber. Whether in any circumstances it would be the plain duty of a scientific man to refuse Government employment because he foresaw that inhuman use would be made of his researches, is one of those neat problems which moral theology loves to discuss. In the case of the atomic bomb, it is clear that the responsibility for using it, instead of holding it as a threat (like poison gas) to deter others from using it, lies not with men of science but with statesmen, who may be trusted to criticize one another without any help from outside. Pulpit orators, it is to be hoped, will resist the temptation to identify science as the enemy of mankind; as if that would discredit the findings of modern physics, and as if you could dispose of an intellectual case by blackguarding the plaintiff's attorney.

It is plain that if the conditions of the modern world allow us to live at all, we shall live to see astonishing progress in this, and perhaps in other fields of science, which will leave us honestly puzzled, and more than a little irritated by the dogmatism of our juniors. There will be moments at which it will look as if the ground had been cut away from all our certainties, and we shall wish for the sake of our own peace of mind, what many people are already wishing for the sake of the World's peace, that the atom had been left unsplit. What is important is that we should not enter upon such an epoch in a doubt-conditioned frame of mind, ready to give up the struggle of thinking, and with it, perhaps,—for in these days an unsupernatural code of morals limps perilously—the struggle of living. A fate with which the modern age is sometimes threatened by its prophets is that of becoming enslaved to its own machinery. But the remedy lies with ourselves; we only capitulate to the machine if we allow suggestion from without to produce in us a mechanized habit of mind.

ANALYSIS

II

Minutes of a Debate

A SENSE of history, a sense of language—these are coming to be regarded as luxurious extras in the curriculum of education; hardly to be acquired in the intervals of that practical schooling which, before long, will qualify every citizen to mend his own wireless set. But these humane accomplishments, which help us so little when we try to master a newspaper article on radio-activity, do give us stray side-lights on the present day which are not without their value. The sense of language enables us, when we pick up a book, to find out exactly what it is our author is saying or trying to say, occasionally with a clearer idea of it than he has himself. The sense of history enables us to see his views, and the importance of his views, as belonging to a period, and conditioned by the thought of a period; to see Man's ideas, even when they dazzle us most by their up-to-dateness, as part of a pattern which is still in the weaving. We detect differences and resemblances between the modern and the ancient world which help us to take the long view, save us from overhasty enthusiasms and despairs.

The very word "atom" is a case in point. As we have seen, it is a term invented to designate something which it no longer designates, the irreducible minimum of matter. At the same time, you may almost say that it was invented to produce, what I was suggesting in the last chapter it threatens to produce nowadays—a doubt as to the spiritual origin of the universe and the existence of an intelligent Creator. Not, indeed, that the debates of early Greece were precisely theological. The Greeks, with their curiously detached attitude towards religion, shunted theology into a siding when they fell to discussing the origin of things. The most reverent of their transcendentalists refrained from discussing what relation there was, if any, between the Supreme Good or the First Cause of their philosophy and that Olympian Zeus whom their fathers had taught them to worship. Their most convinced materialists, the school of Epicurus, granted to the gods of popular tradition a contemptuous permission to exist, so long as they did not interfere; they were sleeping partners in a cosmos they had done nothing to create, constitutional monarchs in a world over which they exercised no control. The debate, therefore, was not one between theist and atheist, rather one between spiritualist and materialist. The introvert projected his own experience into the world around him, and declared (for example) that the cosmos was governed by Love and Hate; the extrovert stubbornly asserted that all was Water or all was Fire, trying to unify his experience of outward things in terms of nothing but what he found there. Democritus, with no laboratory experiment to aid him, developed the atomic theory of Leucippus almost explicitly to justify materialism. The universe consisted of infinitesimal particles of matter streaming through void space, grouping as they did so to form the substances we know. Tiny as they were, these particles had shape, and clung together or parted by the law of their own angularity. A swerve or bias in the current of the stream, obscurely conceived and still more obscurely accounted for, was postulated to explain their continual regrouping.

Democritus is long dead, and if his system has a living appeal for us, it is due to a curious accident for which he was in no way responsible. The doctrines he had put abroad came down, by way of Epicurus, to the Roman poet Lucretius; a man to whom, as perhaps to no one else, we might pay the curious compliment of saying that he had a genius for irreligion. At the very threshold of the Christian era, he sounded the Last Post of an unredeemed world in the haunting phrases of his *De Rerum Natura*; so full of humanity, so much of kin to the materialists of our own day, that he bridges the centuries. He celebrates the

Atom as Pascal, but for the grace of God, might have cel-
ebrated it; hymns its melancholy flitting through the void
as another man might have hymned the story of a great
deliverance. But philosophically he left it where it was, a
lifeless lump of matter, eternally clasping and unclasping
its invisible fellows in the toils of a meaningless embrace.

Meanwhile, science had been in the making, with little debt to the philosophers. Those who blame the Middle
Ages for their want of practical research would do well to
consider whether their quarrel is not really with the men
of a much earlier world. If it is true, as there is good reason to think it is true, that the Alexandrians were on the
way to discovering the microscope and the steam-engine
a little after, if not actually before, the Christian era, what
curious chance was it that killed a tradition, not to come to
life again for a millennium? We forget how much science
depends on its machine-tools; and with these weapons in
its hand, what might it not have done? Yet the work of Euclid and Archimedes somehow fell by the wayside. Caesar
must conquer Gaul without tanks, and the Christian message must come into the world to find it corrupt indeed,
but not matter-conditioned.

On the contrary, during the early centuries there is
a marked tendency, not so much in Christian thought as
in the thought of rival religious systems, away from external reality, away from matter. The Gnostics, and after

them the Manichees, treat the whole visible order as an awkward fact that has somehow got to be explained away; complicated hierarchies of angels and demiurges must be invoked to account for the regrettable accident by which Man came to have a body, or that body to have gross needs and an external environment—one thing was certain, God could not be responsible for such an act of miscreation. The Neo-Platonists, pushing the ideal system of their master to extremes, would fain live in a world of universals altogether divorced from particular facts and particular instances, so disconcerting to the intellectual approach. Christianity, in that first age, far from having to conduct a polemical campaign against materialism, far from having to provide an escape for minds jaded by overmuch contact with the hard realities of life, was for ever having to assert stubbornly that matter was there, and was not evil. Nay, by some strange chain of circumstances never properly accounted for, the Manichean heresy, long dormant in the East, broke out again in Europe in the eleventh and twelfth centuries, so that once more it was the business of Christian thinkers to uphold the claims of matter as something not to be despised.

But admittedly, during the Dark Ages, the rare intellectual life of Christendom had taken St. Augustine for its model, and St. Augustine was impenitently a Platonist. There was obviously much in common between the language of St. John's Gospel and the language of the Dialogues; nor was it a chance to be missed, when you were trying to tame the baser passions of a barbarian world, and a system lay ready to your hand which gave out that all sublunary things were of their nature inadequate and transient. The dominant thought of Europe was certainly Platonist when the schoolmen took control of it. What the moderns usually forget to give them credit for, is having dragged the world back from Plato to Aristotle. To hear the moderns talk, you would think that St. Thomas was a die-hard Conservative who could not rid his mind of the old Aristotelian way of thinking. Actually he was a daring innovator, who risked the charge of heresy in recalling us from a philosophy based on notions of the mind to a philosophy based on experience. It needed the more courage, because Aristotle had been brought back from the East by Saracen conquerors, translated into Arabic and garbled at that. Yet this extraordinary man, in a short life of incessant literary activity, constructed and imposed on his generation a synthesis between philosophy and religion which his Arabian rivals failed, in the end, to secure for themselves. Christianity entered on a new world of speculation, while an Eastern rigidity numbed, and permanently, the thought of Islam.

But it was a ready-made philosophy, not a tradition of research, that had been rescued from the ruins of Greek

civilization. It was the task of St. Thomas to make a Christian of Aristotle, not to make a better scientist of him. St.
Albert, who helped in the process, was indeed a research
student, no less renowned than Roger Bacon himself. But
it was St. Thomas who became, to the generations which
followed, the model of what a philosopher should be, and
research was left to the alchemists, whom the popular
mind obstinately associated with magic. If it be asked what
it was the scientists of the later Middle Ages were working
on, perhaps the simplest answer is, that they were working
on the Atom. The Aristotelian philosophy had much to say
about the *materia prima*, matter still undifferentiated by
form, the substratum of all the different forms in nature. If
this common substratum existed, the alchemist reflected,
it should be possible to turn, say, one metal into another;
you had only to get down to the undifferentiated heart of
them…. Research, at all times, has depended to some extent upon endowment, and it was not difficult to foresee
what kind of research a medieval monarch would be inclined to endow in the circumstances. They set the alchemists to work at turning base metal into gold. The belief in
this possibility lingered as late as Newton's time; then we
all fell to laughing heartily at the stupid medieval fellows
for imagining that you could turn one metal into another.
Today, we are not so certain; the dream of the alchemists
might come true, some fine morning, in our more fully

equipped laboratories; it would be interesting to see with what effect on our economic system. And if we still feel inclined to criticize those old kings for their mercenary attitude towards research, let us chasten ourselves by reflecting how we did at last split the atom, and what use we made of it.

There were by-products of these activities; gunpowder, spectacles, printing are all the machine-tools of science, and we do less than justice to the medieval world if we do not acknowledge our debt to it. But in the main, it must be confessed that the best intellects of the time gave themselves up to abstract speculation, which shows, as the centuries proceed, a law of diminishing returns. It is unfair to criticize the schoolmen for their indifference to the inductive process, unwise to defend them on the ground that they had minds too lofty to be content with grovelling among the data of sense. The desire of the human mind to know is a spontaneous impulse which should be allowed to find its own channel, varying according to the fashion of the age. You should not cramp its genius by digging out irrigation channels for it and bidding it turn the mill-wheels of civilization; solve your currency problems by making gold, or blast your enemies with wholesale destruction. The schoolmen argued about what interested them; it was their right. If their speculations did not lead them to discover the circulation of the blood, or the transit of Venus, there is no agreed standard of values by which they can be called to account for it.

Our loss is that they couched the eternal verities in language which was then the jargon of the laboratory, and is the jargon of the laboratory no longer. Nor does the language of common life stand still; if you live in a culture which has grown away from the scholastic tradition, to learn your theology means learning a new vocabulary, means using familiar terms in an unfamiliar sense. The language of formal theology has become something hieratic, like the language or the accoutrements of devotion. This is disconcerting, when you are dealing with a system of thought which, like the Aristotelian system of thought, explicitly takes its point of departure from common life. Our metaphysical principles might be expected to emerge from our study of physics; but the student who should digest a modern manual of physics by way of preparing himself for a course of St. Thomas would do worse than waste his labour; his preparation, instead of throwing light on his author, would only intensify darkness. An accident of history has put us all at cross-purposes.

Nowhere is this inconvenience more observable than where it is most vital; namely, when the schoolmen set out to convince us about the fact of God. They invite us to apply to the interpretation of existence as a whole those familiar principles which we apply to the interpretation of its parts. But those principles no longer ring with a familiar sound. They are meant to fit in naturally with our alehouse debates; but for most of us, to open a book of formal apologetic is to step into a cool, remote cloister. The ladder that is meant to climb heaven from our front door-step climbs it, instead, from a period world which only history recaptures for us. It is definitely pre-Atomic.

During that astonishing efflorescence of research that marks the seventeenth century, science and metaphysics drew farther than ever apart. "*I* must exist, or I should not be in a position to consider whether I exist or not"—with one phrase, Descartes would cut philosophy and theology away from all contact with rude material things; would have us evolve both from the contemplation of ourselves as thinking beings. He made absolute, by a decree which has lasted to our own day, the divorce between study of the world outside us and study of the human mind as an instrument. (Curiously, for more than a century science made use of its married name, and gave itself out to the world as "experimental philosophy.") Are we to imagine that Descartes was afraid of the new scientific temper, thought it would interfere fatally with scholastic certitude about religion, and determined to raise theological debate beyond the reach of it? It is a much more probable account of him to say that he was the child of his age, and his philosophy represents the great self-absorption which fell upon mankind with the humanist movement. You see it in art, the human figures standing out and the background falling away; you see it in literature, the rise of the drama, ever preoccupied with human situations; you see it in religion, the break-away of Protestantism from the Church-idea; you see it in theology, the introduction of casuistry. To befit such an age, a philosopher must be one who studies, not the thing known, but the mind that knows it. Descartes neglected the apologetic of an earlier day because he was lost in contemplating the greatness of the human mind, just as Pascal neglected it because he was lost in contemplating the miseries of the human soul.

The effect of Descartes upon philosophy, or rather on the estimation in which philosophy is held by the general public, has been fatal. The whole Idealist approach assumes that the human mind is an instrument you can be certain of, and that it does not really make much difference whether you can be certain of the external world or not; thought can go on indefinitely spinning its own web, like the spider with its own stomach for work-box, never acknowledging any debt to brute matter. But, as might have been foreseen, other philosophers arose who questioned the validity of these purely mental intuitions. The same hesitations about the reality of our sensible experience which convinced Berkeley that there must be a God from whom our ideas were borrowed, convinced Hume that you could have no certainty about anything at all. At last, with the air of a ponderous Chief Justice summing up the pros and cons of a suit which had dragged on long enough, Immanuel Kant intervened. He thought he had solved the problem by banishing the categories of Aristotle, and with them the whole fabric of scholasticism, from the soil of reality; meanwhile, he told us to go on believing in God's existence, as a necessary postulate of the human conscience. Whereupon the great heart of the public decided to set down metaphysics as a science at once voluminous and obscure, having no value for the ordinary man, but only for cloudy-minded Germans; an impression which prevails to this day.

And indeed, by this time Science had begun to play its part—the part which is now so formidable—in conditioning the outlook of the plain man. The discoveries of Newton had impressed the imagination with a sense of eternal, unalterable laws governing the whole of existence. Before long, the age of machinery began; and the droves of human beings who now found their employment not in craftsmanship but in turning handles were ready as wax to receive that impression; existence was a great loom which wove its pattern without calling on us for any assistance. It was the age of Deism; you must admit the evidences of a Mind responsible for introducing law and order into the universe, you must rule out miracles, as implying a sort of levity or capriciousness at the heart of things. It was the age of Iron; Existence was a great workshop, full of wheels that turned on their own axles uncomprehendingly, unalterably, without friction, while the absentee Owner of it all lived far away, in a seclusion hardly less remote than the pasteboard heaven of Lucretius.

One thing was left to us—the argument from Design in nature. We could not, like the medieval, find assurance of God's existence in every leaf that budded, every flame that flickered; but surely the marvellous order that existed in the brute universe spoke of a Mind that had planned it? In that, at least, Friar Thomas had not been wrong. Indeed, we were tempted to go a little beyond our book, and infer the action not merely of a Divine Wisdom but of a Divine Benevolence. The instinctive shifts by which unreasoning brutes found their food and secured the propagation of their kind proved that their continued existence was somehow useful to Somebody—Utilitarianism was all about us, and we could hardly doubt that. The very multiplicity of the animal world—Buffon and Cuvier were our authorities for it—seemed to argue a superhuman ingenuity of inventiveness. On one side, at least, science had not let us down.

Then, as if science were determined to let us down on every side, came the *Origin of Species*. The fact that such and such a form of animal life survived did not prove that its existence was a worth-while thing, to us or to Anybody; it had just happened to survive because it was better equipped for surviving. If the world was full of species beautifully adapted to maintain themselves in their surroundings, that did not mean that Providence had been at work; it only meant that the other ones had got eaten. The almost infinite variety of types in nature was due to an age-long process of gradual evolution, during which one characteristic was bred in and another bred out; there was no reason to use the word "marvellous" about it. So died the last of our certainties, as far as creation was concerned. It died almost unnoticed; our theologians were too preoccupied in saving the credit of Genesis to bestow much thought upon it.

At the same time, the public mind underwent a fresh process of conditioning; all mechanism yesterday, it was all evolution today. And evolution was a category under which you could rearrange every department of experience; civilization was evolving, freedom was evolving, thought was evolving, religion was evolving, there was no limit to it. Instead of the endless wheels that used to turn in our heads when we could not get to sleep, nightmare figures succeeded; "dragons of the prime that tore each other in their slime," whose struggle to exist would last eternally; all our struggles were only the continuation of theirs. Instead of the dreary confidence that nothing ever

changed, we had a dreary confidence that nothing ever remained the same; we were back at Heraclitus' difficulty; all
was flux, and we were part of it. The philosophers spread
their ragged sails in the hope of a breeze, but metaphysics,
the science of Being, had less chance than ever in a world
that was all Becoming.

We thought we had reached the last pitch of self-abasement, but we were evolving still. By now science had grown
so specialized that the man of ordinary intelligence could
not hope to follow what it was doing, or even what it was
saying, with any comprehension; least of all if he was un-
trained in higher mathematics. Then Einstein dawned on
the horizon, and all horizons vanished. A new notion had
come to condition our minds—this time, Relativity; bred
in a culture of terms we could not understand, and calculations we could not follow. Hitherto, although we were
robbed of Aristotle, Euclid had been left us; now, not even
the basic certitudes of the mind, certitudes which even
Kant had tolerated in the mind, as long as they did not
seek to go outside it, were hauled up for revision. I have
never heard an exponent of Relativity deny the principle
that "things which are equal to the same thing are equal to
one another," but none of us would be surprised at being
told that "it only holds good at a certain level." And relativity threatened to be a form of mind-conditioning even
more relentless than evolution. Evolution in morals was

bad enough; where should we be when it came to relativity in morals? The stars in their courses were fighting against us, if we could still believe that they had any.

One hope remained to us; it had almost begun to look as if science was going the same way as philosophy. Descartes had cut the ground from under his own feet by divorcing metaphysics from reality; what if Einstein and his followers should carry off science into the empyrean of pure mathematics, and leave us others to grovel in our old, common-sense certitudes once more? For indeed, the language the physicists were beginning to talk about the composition of matter was hardly less confusing than even Einstein; telescope and microscope alike seemed to have outshot their range, outlived their usefulness, and given on to a looking-glass world which had no real connection with this our terrestrial existence. All very well to talk about the atom; but had it any real importance, except for the theorist? We felt inclined to dismiss the whole subject by going out and kicking a large stone, like Dr. Johnson refuting Berkeley. Was the atom really going to come into our lives?

And the answer came, more terrible than thunder, from Hiroshima.

III

Thomism and Atomism

No ATTEMPT shall be made here to examine the relations of philosophy and science in general, or the relations
of scholastic philosophy and modern physics in particular,
on the level of academic debate. Let others, better qualified, face the abstract problem; let us have a stop-press
edition of Père Garrigou-Lagrange's *Dieu, son Existence et
sa Nature*. Everything is in a mind, my business is with
the plain man, and his reactions to the confused portents of our time. "Happy, who Nature's riddle learned to
read," but that part is forbidden to me, as to Virgil, by the
slow coursing of the blood round my brain-cells (I imagine that is what Virgil meant). "River and wood afford my
humbler theme"—let us be content to construct an asses'
bridge over the difficulty, to find some twilit path through
the dark forest of controversy, as best we may. What is it
that gets us down when we hear, almost in adjacent lecture-rooms, the metaphysician talking *his* native language,

and the modern scientist talking *his*? Whence arises the uneasy suspicion that our minds are being conditioned in two different ways at once; this way when we go to Church, that way when we pick up the morning paper?

Perhaps at the back of our minds—and it is that mental background which concerns us—there is a sense of disappointment that science and metaphysics have gone on all these centuries, each trying in its own way to analyse human experience, and never seeming to travel by the same road; nay, if anything, straying farther apart. Truth is one; we cannot doubt the sterling quality of that philosophic system which is so deeply encrusted in the precious fabric of theology; on the other side, can we suspect mere paste-board glitter in the brilliant science of our day, which opens for us so many windows of experience? And yet they never seem to match. To put the problem at its most infantile; how (we ask) is it possible for research to burrow deeper and deeper into the very heart of being, and come back to us with no news of having come across, even having got nearer, the heart of being, as philosophers conceive it? We talk about "form" and "matter"; distinguish between the mere undifferentiated substratum which underlies any existing thing, and the added principle which makes it what it is. And here are the physicists, splitting up the molecule into atoms and now picking away at the atom itself, peering down into a deep abyss in which the constituent elements of all chemical things are the same; yet never a word have they to tell us about where "form" ends and "matter" begins! We are still more familiar with the distinction between "substance" and "accidents"; all the qualities in a thing on which our senses report to us, ending with the termination "-ness," the blackness, sweetness, thickness, for example, which mark out the perfect coffee, are only accidents which "inhere" in the substance of the thing in front of us. Depending as we do on the senses for our information, we could never (we were told) come in contact with the substance itself in our daily experience; *that* always eluded our senses, but we knew it must be there. Was it possible, we asked ourselves, that the daring enterprise of the modern laboratory would do something to clear up the ancient riddle for us? But no; protons, electrons, nuclei are there if we want them, but of substance divorced from its accidents never a word.

That, as I say, is to put the problem at its most infantile. We knew really that formless matter and substance without accidents were not to be found any more at one end of a microscope than at the other; by definition they must elude our most searching enquiry, if it was, in the last resort, our senses that had to be called in evidence. The scientist and the metaphysician are not taking two divergent roads in an attempt to run down the same quarry; they are barking up two different trees. The professor who offered to show us a piece of formless matter in a test-tube would obviously be a fraud. True, he may write out a formula on a piece of paper, giving you the mathematical representation of such and such forces, themselves beyond the reach of the calliper and the scales, which underlie such and such a material substance. But the formula is not the substance, any more than an index number on a card is a colour. The scientist is representing reality to himself in mathematical terms, while the philosopher represents them to himself in metaphysical terms; they do not meet, and there is no reason why they should.

What lies, I suspect, at the root of our discomposure is, that when we sit down to philosophize about the world of our experience we instinctively represent it to ourselves in good, old-fashioned, materialist style; we think of the world as a set of lumps. We set before the mind's eye a lump, say, of coal, and remind ourselves that the blackness of it is not the coal, the squareness of it is not the coal, and so on; the very solidity of the object in front of us acts as a kind of fulcrum against which the agility of the intellect can brace itself. If the scientist comes along, and tells us that what we are looking at is not really a solid lump, it is a stream of whirling electrons, we are thrown out of our stride. We are trying to abstract from the aspects of the thing which fall under the observation of the senses; and whether or no it be a stream of whirling electrons, it is quite certain that a stream of whirling electrons cannot fall under the observation of the senses. Let us have solid bodies to philosophize over, not a kind of ectoplasm.

Entities, the schoolmen said—and in that, at least, they were surely right—should not be unnecessarily multiplied. And we are tempted to feel there is something uneconomic about having three separate interpretations of the external world and juggling with all three simultaneously; the solid world of common sense, the world of the metaphysician, all docketed and labelled like an album with pressed flowers in it, and the world of the physicist, rotating, coruscating, ebullient with paradox. How (we are tempted to wonder) does an angelic intelligence see our lump of coal? Not lump-wise, we may be sure of that; such a view of it could only present itself to our terrestrial slowness of wit. But does the angel see a bunch of accidents inhering in a substance, or a stream of whirling electrons, or both, or something yet other, beyond all these? The profane reader must not be impatient with me for bringing angels upon the scene, as if I were trying to beg the question of theology. We are simply concerned to answer the age-long riddle, What is real reality really like? And the new guess of the physicists, far from settling the problem once for all, proves a fresh source of bewilderment. We cannot believe it is the whole explanation; it is too full of dots and dashes for that. Yet it cannot be a pure chimera of the intellect; there is, Hiroshima knows, a dreadful reality about it. How are we to integrate our worldview, with this double astigmatism obscuring it?

Meanwhile, none of us likes to be old-fashioned. It is the newest song—Homer assured us of it, centuries ago—that ever sounds most gratefully in men's ears.

There is something dated, surely, about this talk of natures, and forms, and essences; our minds misgive us lest they should be a mere ornament, not a weapon of our thought, comparable to those brightly-polished warming-pans that hang, unused, on the walls of an old inn that has been "done up." At best, like an old stoup now used for an ash-tray. After all, if we had been living in the thirteenth century, we should have found the philosopher in his cell and the alchemist in his hide-out talking the same language; using the same terms, "nature," "form" and "essence," with the same meaning. St. Albert, pegging away at his botany, and St. Thomas, discussing whether or no the angels could be divided into species, were fast allies, and lived by a common culture; their modern representatives move in different orbits. In an age that has an itch for modernity, difficult not to feel that our *philosophia perennis*, however much wear may be left in it, has had the nap rubbed off it by time.

And this, even when we are imparting to one another the immemorial objections, the immemorial come-backs, in the privacy of our own lecture-rooms. Worse still, when we must go out into the open, and discuss the fallacy of the infinite regress before the shifting crowd that eddies round Marble Arch. For, after all, people have got to believe that God exists before they will consult Church or Bible to find out more about him. And the main proofs of his existence given in our textbooks are still the proofs which St. Thomas gave us; a few more defensive earthworks have been thrown up, but the position is still the same. We are expected to prove God's existence by pointing to the natural world and inferring from it, without the possibility of error, the presence in the background of a supernatural Coefficient; or rather, Efficient.

You are traveling by a local train; with no excuse of halt or station, it seems to lose interest and pulls up. After a little clucking and scowling among your fellow-passengers, one of them cranes his neck out of the window and returns big with information; "Signals against us," he says. With that, the whole party is content, and sinks back into British tolerance of the *fait accompli*. But in fact the explanation he has offered is only the broken fragment of an explanation. The signal is up because the train in front of you has not yet cleared the signals; why not? They are still awaiting, perhaps, a fast train on the main line which has the priority. And why is the fast train late? A further delay must be postulated to account for that; and so on and so on—but not *ad infinitum*. Even if the number of trains were infinite, an infinite number of trains held up by signals would give us no account of the matter; would not tell us why any signals are up at all. Somewhere, it may be two hundred miles away, something *real* has happened; there has been an accident, or a carriage has caught fire, or an unexpectedly heavy consignment of luggage has made it impossible for a guard to blow his whistle in time, or a piece of line is being mended. You have not accounted for the hold-up of your own train until you have traced the reason of it, signal-box by signal-box, to this *real* interference with the scheme of railway punctuality. I seem to have heard that either the ten o'clock train from York to King's Cross, or the ten o'clock train from King's Cross to York, is the daily foundation of all our railway service; if that train starts late, it means that before long all the trains will be behind schedule, all over the country.

The above parable is offered in the hope that it will make the argument from causality somewhat more lucid than the illustrations commonly given in the textbooks. True, the causes postulated to explain our railway embarrassment, are, for the most part, what the philosopher would call occasional, what the theologian would call intentional causes. But it will serve to explain what is often insufficiently explained in the presentation of the argument; namely, why you cannot get out of it by postulating an infinite series of causes. My train is being held up by a signal on account of another train which is being held up by another signal, and so on and so on and so on; but no sensible person would imagine that by putting the responsibility always one stage farther back you could avoid, ultimately, the conclusion that something, somewhere, has gone wrong. So it must be, the scholastic argues, with physical causes; the kind of causality they possess is of its own nature inconclusive; when you have traced a rise in the mortality statistics to an outbreak of typhoid, and the outbreak of typhoid to a bad water supply, and the bad water supply to an injudicious disposition of the local drains, you are plainly looking down a long vista of explanations which lead you nowhere. Or rather, for practical purposes of hygiene it may be of supreme importance, but it is never going to satisfy the mind with a sense of ultimate explanation.

Most of us, when we were first introduced to the science of apologetics, took up this point all wrong. We modelled our half-formed notion of the argument from causality on a string of trucks being shunted by an engine; as if creation consisted of a multitude of hard objects each of which was being pushed by the one behind, the ultimate impulse, miles away and centuries ago, being contributed by an Omnipotent Creator. And this notion has two obvious defects. It pictures God's dealings with Creation as

Deism pictured them; the universe becomes a complicated machine which an Almighty Craftsman has devised and set in motion, *and thereupon left alone*. And it also means thinking of the First Cause as something *within the material order*, distinguished from other causes in that its activity is self-originated, but "causing" things in exactly the same sense. It is quite evident that St. Thomas did not mean us to understand anything of the kind, for he distinguished God as the Primary Cause from those secondary causes to which we attribute this or that effect in our daily experience, and taught that the influence of the Primary Cause was present everywhere, conspiring with the secondary cause to produce the effect. In following up the clue of causality which the material order gave us, we have passed beyond the material order, into a region where our ideas about hard objects pushing one another no longer apply.

Still, it *was* from our experience of the created world that we took our point of departure; it was that experience that enabled us to form all our notions of causality. There were, to be sure, two other ways of approaching the subject. You could point to the existence of change in the material world, and challenge your opponent to maintain that the agent in the change was either the thing which was there before, or the thing which was there after, the change had taken place. Or you could plead that a whole universe of individual things, each of which showed by its impermanence that it had (so to speak) no right to be there, demanded a background of Necessary Being as the sufficient explanation of all this contingency. But, although you learnt how to distinguish these two methods of argument from the one already mentioned, they both looked uncommonly like an argument from causality, and the word "cause" was seldom kept out of the text by the author with whom you were discussing them. In the long run, you felt, the first three Proofs stood or fell by the value of the causality argument.

That its value had been disputed, we knew; the text-books told us about it. Idealists and Sensationalists had conspired to assure us that causality had no roots in the external order of things; it was an *a priori* notion which the mind brought with it to its interpretation of reality; the feast of reason to which philosophy invited us was (if the crudeness of the metaphor may be allowed) only a bottle-party. The Sensationalist's conclusion was, "So much the worse for the Mind," while the Idealist was content to say, "So much the worse for external reality." But they were agreed in thinking us wrong; were we impressed at all by their criticisms? Probably not; the Sensationalist could not explain how or why the notion of invariable sequence gave rise to, and passed into, that of causality; the Idealist could establish no probability that a set of innate ideas, owing nothing to any contact with reality, flourished in the minds of the newly-born. But then, while these philosophies were in their hey-day, before Queen Victoria came to the throne, there was a good plain man's argument to demolish them with—if causality was a mere thing of the mind, how was it that science, by patiently tracing the *causes* of things, had discovered the circulation of the blood, and given us the steam-engine?

Now, by a most disconcerting *volte-face*; our old allies have deserted us. The scientists tell us in so many words that they do not bother about the causes of things any longer; the very word has passed out of their vocabulary. When they try to explain what conception of the natural world has taken its place, they are no longer so successful in conveying (to me, at any rate) their precise meaning. I think their complaint is this—that our world-picture is built up on the notion of one thing influencing another, acting upon another (as when we say, "Acids turn litmus paper red"), whereas science thinks of them as interacting with one another. After all, while the fire makes the water in our kettle hot, the water is all the time making the fire cooler. Or rather, perhaps I ought to say, the temperature is being redistributed; we must learn not to personify, if we are to avoid getting into trouble with the scientists. Anyhow, they want us to stop thinking of the material world as a collection of things pushing one another about, and think of it rather as a pattern whose threads are interwoven in a discoverable way.

I have used the word, "personify"—perhaps not the least of the worries we are apt to feel over the scholastic account of the universe is just that; did they, did Aristotle, perhaps import too much of our human experience, the experience of thinking beings, into their interpretation of brute nature? The personifying tendency is, after all, an ascertainable fact. I can remember myself, when I had fallen out of a trap as a very small boy, bringing out a whip and lashing the gravel on which I had fallen, pretending, by a conscious fiction, that it was morally responsible for the abrasions on my hands and knees. Nor was I the first to act in this way, though I doubt if I knew, or at any rate remembered, how Xerxes had lashed the waves of the Hellespont. *Omne agens agit sibi simile*, St. Thomas writes, but can the verb "to act," strictly speaking, have its subject in the neuter? Does not the very word imply purposive action, under the guidance of a self-contained will? Not that St. Thomas is any more to blame than the science of yesterday (if not of today) which used to talk about "agents" in a purely chemical context. I dare say the distinguished men who worked out the atomic bomb, cut off from our profane audience by the barbed wire in which a prudent Government had isolated them, used the word "cause" more than once without thinking. But all our language is pictorial, and we must not build too much on the use of a word. The point is, Did the schoolmen read into their interpretation of the outside world notions which really belong only to the self-conscious experience of human agents? And if so, was "cause" perhaps one of them?

I do not ask that question with the intention of suggesting an answer, in this chapter anyhow. I am only trying to drag out into the light those secret misgivings which assail us, or perhaps I ought to say that mood of *malaise* which afflicts us, when we put down the *Summa* and take up the *Proceedings of the Royal Society*. Pragmatism is always tugging at the outskirts of the mind, and just as it was encouraging, a century or more ago, to feel that the great discoveries of the laboratory were being made in the name of cause and effect, like our own leading proof of religion, so it depresses us, nowadays, to feel that physicists are dispensing, in their triumphant progress, with the very machine-tools of our thought. Regretfully we put aside (without for a moment consenting to discard) the argument from causality, and with it the two other arguments which have such a strong family resemblance to it. There are two proofs still remaining, one of which is so reminiscent of Plato that you feel it is hardly fair to plead it in defence of Aristotle, and the other—ah, the other, that at least is so strongly engrained in our way of thinking that it will take more than a little neglect on the part of scientists to make us distrustful of it. The proof from order, from the adaptation of means to ends on a level of existence where *conscious* adaptation of means to ends is out of the question; how does that emerge, when it has passed through the crucible of our modern alchemy?

The difficulty, as I see it, of keeping this argument in the forefront of our minds is the difficulty of continuing to be surprised at a thing, in itself astonishing, which is a matter of daily experience, which is so familiar to us that we take it for granted like the air we breathe. Here at least we are not wrong if we import our own human experience into the picture. It is essentially because we are reasoning, purposive beings that we claim to be able to judge, whether the activities of a reasoning, purposive Being are traceable in the world of our environment; the Artist (may he forgive us the boast!) can only be appreciated by his fellow-artists. In the nature of things, why should we expect to find any system or symmetry in the nature of things? We all know the kind of Englishman who is disgusted, when he travels abroad, to find that these foreigners drive their cars on the wrong side of the road and put their verbs at the wrong end of the sentence—everything is "all anyhow." Why does not Man find himself equally puzzled by his natural surroundings, if the cosmos is really a chaos, *if* there is no uniformity there apart from the ideas of uniformity which his own mind brings in with it?

We might be inclined to answer that Man has got accustomed to his surroundings by now; he is like the child born and bred in a squalid slum, that has come to expect nothing but squalor. It will not do; you cannot expect the eternally variable. When people talk as if the uniformity of nature was really a uniformity of our minds which we have contrived to read into nature, I am always reminded of a picture I saw long ago in a cheap magazine. It represented the edge of a razor-blade, powerfully magnified, looking like a series of jagged peaks, and, side by side with it, the edge of a blade of grass, equally magnified, looking like the edge of a blade of grass. In all honesty, must we not regard our notions of regularity as something we get out of our observation of nature, rather than as something we put into it? The clock is the offspring of the sun-dial, not the sun-dial of the clock.

To be sure, our favourite way of illustrating the uniformity of nature was, in old days, to point out how skilfully the birds built their nests, the beaver its dam, how well qualified for flight were the animals that were ill qualified for self-defence. And the dogma of natural selection gave us pause; the unconscious picture in our minds was no longer that of Nature as a mother wisely protecting her favoured children, but of Nature as a conscientious examiner, ruthlessly eliminating the candidates who do not come up to scratch. But we knew that the principle of natural selection could not really work without other principles to help it out; and the evidences of order in the outside world were too multitudinous to leave us in any habitual doubt. Our eighteenth century mind-conditioning, based on machinery, was not successfully obliterated by our nineteenth-century mind-conditioning, based on the struggle for existence; the universe might be a riddle, but the very fact that it was so obviously a riddle encouraged us to believe in a Mind which had been ingenious enough to set it.

And here, in a curious way, our faith was supported by the unbelief of the unbeliever. Were you tempted to doubt, for a moment, that the natural world was subject to a reign of law, you had only to start an argument about miracles with the first person you met, and the overwhelming conviction of the human mind in favour of a cosmos determined by unalterable principles became evident to you.

> "There is one creed; 'neath no world-terror's wing
>   Apples forget to grow on apple-trees"—

we knew little about world-terrors when Chesterton wrote the lines, but they came back to us in the dark days of Mons and Ypres, steadying us with the chilly assurance that the world of things still answered its helm, at a time when the world of humanity seemed to have gone hopelessly adrift.

RONALD KNOX

And now, when two wars lie behind us, and the current of human affairs seems swifter and more ungovernable than ever, we turn despairingly to the scientists who have presented us with the Atom Bomb, and ask, at least, to have their word for it that this cataclysm of history is not a cataclysm in the whole structure of our sublunary existence. "Tell us at least," we insist, "that you have looked through your microscopes into the very heart of nature, and that order still reigns *there*."

"Well," they answer, "we are not quite sure. The days have gone by when freedom of research was fettered by preconceived ideas about the principles on which nature is organized; when Galileo was chidden for doubting the immobility of the heavenly bodies, when Père Noel would fain have demonstrated to Pascal the metaphysical impossibility of a vacuum. Not for us to embarrass our successors by trying to lay down immutable laws based on empirical certainty. But if you want to know the state of the case, this is what our present experience seems to indicate. The moment at which a radium atom will explode is a moment which we cannot predict; that is certain. But more; the best opinion is that the moment at which a radium atom will explode is a thing essentially unpredictable; if you had all the knowledge of men and angels, you would not be able to predict it, because it seems to lie at the discretion of mere chance. And if that is so, a kind of anarchy seems to reign in the very heart of nature; the law of averages comes in, no doubt, to redress the balance; but to assert that there are any other laws in nature is to go beyond our present evidence. It looks very much as if indeterminacy were a fact."

And we creep away thanking them, our last certitude abolished.

≫

IV

The Fading of a Dream

"THERE WAS only one shower while we were abroad….
Poor reasoners! who think any instance of Providence too
small to be observed or acknowledged." "The sun shone
extremely hot on my head, but presently a cloud inter-
posed. And when I began to be chill…it removed till I
wanted it again. How easily may we see the hand of God in
things small as well as great! And why should a little point-
less raillery make us ashamed to acknowledge it?" Any
reader of Wesley's Journal knows how commonly such en-
tries occur, how seldom he is content to record the fact of
any preservation from danger, or even from discomfort,
without drawing a moral from it. To him it seemed quite
evident that a special Providence watched over every detail of his career, that sun and rain and wind were attem-
pered, again and again, not merely to his prayers but to
his unspoken wishes. He does not even stop to consider
the difficulty that what was a providential circumstance

for himself might easily be an ill wind for somebody else; he does not weary us with philosophizing about primary and secondary causes. That all things work together for our good, a confidence most of us retain, if we retain it at all, by much exercise of faith and much allowance for unknown factors in the situation, is to him simply a matter of daily experience; he throws the evidence at us, and (since the diarist is free to select his material at will) we are left with nothing to say.

The "reasoning" and "raillery" he alludes to are primarily those of the Deists. He is reacting, you see, against the tendency of his age; he is in revolt against mechanism. He took his M.A. in the year of Newton's death, and himself died in the year when Faraday was born; his attitude towards the sciences is characteristic. "I doubt," he writes, "whether any man on earth knows either the distance or magnitude, I will not say of a fixed star, but of Saturn or Jupiter, yea, of the sun or moon." *There* he is behind the times, because he will not capitulate to the advances of determinism; try him on electricity, and you find an enthusiast at once. "I went with two or three friends to see what are called the Electrical experiments. How must these confound those poor half-thinkers, who will believe nothing but what they can comprehend!" He mistook a rift in the clouds for clearing weather; dared to hope that science, which had all but enslaved our imaginations, was preparing to set them free. It took a lot to condition the mind of John Wesley.

Yet Wesley's terrific career was essentially a protest, a reaction. Humanity had seen a ghost, and it would no longer be possible for us to accept naturally and instinctively, without theological reflection, the doctrine of special Providences. If an itinerant preacher was spared a wetting, *we* should be more ready to congratulate him on the inspiration which had led him to get up in that particular place at that particular hour on that particular day, than on any direct interposition of Omnipotence. Unpredictable rain and sunshine may be, but we no longer cherish the confidence at the back of our minds that they are something undetermined. Our minds are too machine-conditioned, in spite of a hundred disappointments, to doubt the weather report.

But there is another difference which marks us off sharply from that twilight of culture in which Wesley lived. The world is too crowded now, and has been for more than a century past, to let us be contented individualists. You find Wesley making the calculation that if Liverpool goes on growing at its existing rate, it will soon be nearly as large as Bristol.... To be sure, there is no logical reason why Omnipotence should find it more difficult to look after forty million people than four million; but the imagination, discipline it as we may, is daunted by high totals. Our instinct, as the cities cluster thicker about us, is to think more in terms of averages and of movements; the interests of the community, and of ourselves as a part of the community, present themselves readily to the mind. (I can remember, as a small boy, praying for rain so that it might interfere with the cricket, but I do not think I was a typical small boy, or a nice one.) We go on praying for individual benefits, and feeling grateful over them; the instinct is unconquerable, unless you have altogether parted company with religion. But our public attitude, if I may put it in that way, is to hope rather for the wide distribution of good things than for their accurate canalization.

And as the Victorian age went on, this sense that the world was overcrowded became intenser; it was not only our fellow-men, but our fellow-creatures, that were too much with us. Darwinism made us all animal-conscious, by suggesting that we had a common ancestry with the animals; would it not be narrow-minded if we pictured the world as a pleasure-ground meant only for us men, not for the cadet branches of the family as well? After all, there was a Divine munificence which providently catered for the sparrows.... But then, the difficulty arose, where did this process stop? Was the amoeba catered for, or did the amoeba have to look after itself? We recoiled from the *reductio ad absurdum*, and substituted for the picture of Providence busying itself over the individual sparrow the conception of nature as a kindly mother on the whole, ruthless to the individual but "careful of the type." There was a purpose, anyhow, running through the ages; we were not the sport of blind chance; let that suffice for our instinct of self-importance.

You see, where there is belief in God's existence, there must be belief in his assistance as well; hope is the natural food of faith. The God of the Deists was a convenient postulate, he was not really an object of worship. As well expect the religious emotion to be kindled by the shadowy heaven Lucretius pictured for us, as by a God who set Creation afloat, and left it to sink or swim. Nor, west of Suez, will humanity bow the knee to the God of Pantheism, a God caught up in the wheels of his own machine, no more necessary to our existence than we are necessary to his. True, in the higher walks of mysticism you will come across rare souls who aspire to love God only for what he is, not at all in acknowledgement of what he does for us. But the rank and file of worshippers will demand, always, that God should take an interest in his creatures; he must not be ashamed to be called *their* God. Let him punish them sevenfold for their sins; let him hide his face away for a time, so as to quicken their longing for him; let him deal out sorrow as well as joy, to show that he is Master still. But he must have a personal relation to them, somehow, if they are to have a personal relation to him. The heathen may worship gods who have eyes and see not, have ears and hear not, but that is a calf-love which humanity has outgrown; we can never recapture it. The God who reveals himself in complete indifference will have less crowded temples than the hidden, silent God who cares.

Unconsciously, what undermined the religion of the Victorians was not so much that Evolution had emptied heaven for them, as that Evolution had destroyed their confidence in their own individual value; with Nature crying out that a thousand types had gone, was it possible to believe that a single soul, though it were Lord Byron, though it were General Gordon, awoke the Divine interest? Unconsciously, what sustained them more than they knew was, I think, a belief in the destinies of their own country, founded on Waterloo and fostered by the Exhibition at the Crystal Palace. After all, if Nature had her favourite types in the evolution of the animal species, why not in human history as well? And if such favouritism existed, it did not much matter whether you attributed it to "Nature" or to "Providence." The British type seemed to have the marks of a dominant type, qualified for a great Darwinian destiny. It had been fertilized by cross-breeding with Saxon, Dane and Norman. It had maintained itself in a severe struggle for existence, and maintained itself because its amphibious instincts harmonized with its island environment. Its characteristics had hardened in isolation, till you could almost be excused for regarding it as a distinct subspecies of humanity. This half-scientific way of looking at things was fused, easily enough, with a half-religious way of looking at things; the Old Testament was still popular reading, and the notion of a favoured people, bound to its ancestral God by reciprocal obligations, was in the Old Testament manner. If we no longer interpreted our private biographies in the spirit of Wesley, we could still interpret English history in the spirit of Cromwell.

All this was said in an undertone; we were too well bred to say it out loud. Indeed, when Kipling blurted out his confidence in the Divine mission of the race, it was a sign of our senility; a man must be getting near sixty before he begins to boast of his marvellous health. Only those of us, I think, who were born under Queen Victoria can know what it feels like to *assume*, without questioning, that England is permanently top nation; that foreigners do not matter, and if the worst comes to the worst, Lord Salisbury will send a gun-boat…. It suffered on its theoretical side, this confidence of ours, when another European nation, not altogether of our culture, suggested substituting the word "Nordic" for the word "British," and threatened to establish a claim of founder's kin. It suffered on its practical side, when we went off the gold standard. I know that there are plenty of people who still hold it, some of them as a kind of religious creed, some of them merely as a prudent judgement. But they hold it defiantly, in the teeth of objections; they hold it as a dogma of belief, not as a plain fact of observation. The glory has passed, and we are not the sole patentees of the Atomic Bomb.

Meanwhile, another doctrine of Providence was growing up in the background of our thought; or, if it was not a doctrine of Providence, it was at least a compensation for the absence of any such doctrine; I mean, the belief in Progress. As a biological theory, it had less to be said for it than the dogma of British supremacy. Huxley warned us long ago that you could not legitimately regard the moral improvement of the human race (if any) as the continuation of, as being in line with, the development of natural species. In evolution, certain characteristics maintain themselves, and secure the permanence of the breed which possesses them, not because they are something in themselves worth while, but because they are useful weapons in the struggle for existence. That slavery should be abolished, that government should be carried on under democratic forms, that a higher value should be set on human life, that we should be kind to animals—all these things are manifestly desirable in themselves, but it was not by such arts that man attained mastery of the planet. Rather, if anything, the contrary.

Rather, if anything, a certain ruthlessness is needed about the type, if it is to insinuate itself into Nature's good graces. To take the most obvious example, our civilized instinct is to bestow jealous care on the laggards in life's race; to nurse the incurably sick, to bolster up the cripple, to support the aged, to put the old horse out to grass. If we were still, as a race, engaged in the struggle for existence, we should be pursuing just the opposite policy; the iron law of Survival demands that we should let the weakest go to the wall. And, indeed, in these science-ridden days we are half beginning to capitulate; you hear whispers about lethal chambers, and selective breeding. But the Victorians continued to believe that the standards of human life were growing more civilized from decade to decade; their appeal to Darwinism in support of the contention was only a façade, what was really operative in their minds was the old belief in Providence. The same agency which had kept the sun off Wesley's head when he was preaching now watched over the fortunes of the Prohibition movement. It is true, we had grown somewhat Pantheist in our conceptions; this Power making for righteousness was thought of, now, as something residing in us rather than as Someone ruling above us. But it gave us a faith to live by; and those who were religiously inclined found it more and more difficult to disentangle their belief in a supernatural world from their confidence in the bright dawn that awaited the eyes of a happier and a wiser humanity.

The modern world was like Epimetheus, looking into Pandora's box; the disclosures of science had told it enough about its origins, about its limitations, to make it feel unhappy about faith; but hope was left. Those who had lost the sense of religious certainty enrolled themselves under the banner of optimism; the world's future occupied their thoughts, instead of a future world, and, by a kind of inverted Confucianism, they fell to worshipping their grandchildren. With this optimistic agitation, so familiar to us during the last century of our history, the leaders of religion have readily, perhaps too readily, associated themselves. At first, perhaps, from a sense of shame at seeming to lag behindhand while others promoted this or that altruistic movement, but as time went on with more open-hearted enthusiasm. Preoccupation with the individual soul, the urgent sense of its importance and of its peril, had grown rarer in these twilight days, and it was a relief for the generous instincts of the preacher, to have a cause to throw himself into; surely he could not be wrong in devoting himself to the cause of the neglected and of the oppressed? No one who has watched the portents of religion during the last generation will be inclined to dispute that we have moved away, for better or worse, from "otherworldliness"; that many Christians when they pray "thy kingdom come" attach a sense to the petition which belongs to our epoch. We want a kingdom of God on earth, here and now.

It may be observed in passing, that this modern emphasis distinguishes us sharply from the first age of Christianity. In the age of the Fathers, it would seem to have been the common impression that the world-order was soon to reach its end; that the Providence of God was concerned to watch over Christians and rescue them from overmuch alarm in this life, from eternal loss in the next. I am not proposing here to criticize this shifting of emphasis; only to point out that the man who pins his hopes to an improvement of conditions in this world is more exposed to the danger of disappointment here and now than the man who looks forward to a readjustment in the next. He is more ready to be elated or depressed when he opens his morning paper.

The optimists of our immediate time found a cultural centre in Geneva. The years 1914–1918 had produced tragedies on a scale so unparalleled in our earlier experience, that war seemed something utterly barbaric; it was an anachronism, a primeval abuse which we had forgotten to extirpate, and to the task of extirpating it we all devoted ourselves, with a unanimity about the end in view only equalled by our want of agreement about the means for securing it. Somehow, it does not matter how, we mismanaged the affair. At the end of a longer and far more disastrous war, we seem farther than ever from the prospect of a stable peace. The waters of Europe have been deeply stirred, and in many countries the elements which have come to the surface are not the healthiest. We are at the mercy of adventurers, and our modern habit of trying unsuccessful politicians for their lives will hardly, one imagines, make for the emergence of the philosopher-king. Never has the world given thanks for peace with such a deep sense of disillusionment. And those whose trust in Providence was based, more than they knew, on the conviction that the world was steering for Utopia are numbed with bewilderment. For them, as for the hospital-nurse whom I quoted in my first chapter, the conclusion seems to be, "Now I know there is no God."

For all this, it would be unfair to lay the whole blame on the Atom Bomb. If Hiroshima were standing, and Nagasaki as busy a centre of commerce as ever, we should still be facing the future with misgiving. But its detonation forms a kind of signature-tune after all that orgy of destruction which has been going on in the past five years; it is a symbol which has struck the public imagination and deepened its sense of doom. Worse, it is generally accepted as a foretaste of the weapons the next war will be fought with. Weapons whose blast comes round the corner to kill you; weapons that are not content to kill with blast, but leave people with incurable burns on them, impregnate the earth where they fall with the properties of fire…. There is, of course, no reason whatever why a man should not entrust those he loves to God's safe-keeping when these are the dangers they are threatened with, as confidently as in the old, primitive days of the rocket-bomb; for the intellect, the principle is the same. But the imagination *will* make pictures for us, and there are some which hardly bear thinking of. Even if we were only concerned with the safety of our immediate friends, we would give much to be able to banish this kind of nightmare from our dreams.

But, as I have suggested, there are many people nowadays to whom the idea of a special Providence watching over a particular person has become unfamiliar; there are so many difficulties about it, and perhaps they regard the whole notion as rather small-minded. They have, nevertheless, been believing in Providence for years, although they did not call it that; they have been trusting God to see that the world got better and better, so that our grandchildren would look back upon our times as times of inconceivable crudeness, barbarism almost. Once again, there is no reason intellectually why we should not trust Almighty God to tide us over the dangers of the Atom Bomb. Perhaps, as some people hope, the very threat of its horrors will deter people from going to war again ever. Perhaps a fresh discovery will put us in possession of some means by which we can neutralize its effects. Perhaps there will be a convention against its use, and if war does come, either side will be afraid to use it, for fear the other side should use it even more effectively. Perhaps the secret will, after all, be kept, and this formidable power will remain in the hands of peace-loving nations, only to be used, if the mere threat of its use does not suffice, against the unjust aggressor.

All these are possibilities, and it may be that in a few years' time we shall be thinking more calmly about the whole subject. But at the moment, I think, our mood of depression is deepened by the thought of what *might* happen; whole silent cities where none are left to bury the dead, multitudes of people starving in underground shelters because they dare not come out to get food. Worse than that, ideally speaking—the possibility that the latest and most effective use of atomic energy should fall into the hands of some enemy of civilization, as ambitious, as unscrupulous, as ever Nazi Germany was; that the whole world may be enslaved to some evil philosophy, unless it will accept the alternative of annihilation. All our notions of security have been so bound up with prudent human calculations, whether this or that country had the man-power, the capital, the technical skill to become a public danger; this violent short-circuiting of the destructive process throws all our calculations out, gives us the sense that the state of Europe, or the world, is as unpredictable as the radiation of the atom itself.

I suggest, then, that our imaginations are threatened with a break-down of hope, as they are threatened with a break-down of faith; and that if we allow ourselves to brood too much over the idea of an Atomic Age those two assaults upon our religious convictions will get entangled in our minds, each of them imperceptibly coming to the support of the other. And just as we want to think clearly and calmly about modern physical theories, so that they will not jolt us out of our metaphysical convictions, I would urge that we want to think clearly, and as calmly as the nature of the case allows, about the possibilities of disaster that lie before us.

We have to see the greatness of God as a background, not only to the imperfect knowledge we have, and shall always have, about the inner nature of reality; but also as a background to our world-philosophy, always too ready to assume that God's plans are ours, that God's scale of values is a replica of the scale of values we take for granted.

&c.

V

A Missed Opportunity

THIS CHAPTER will be somewhat parenthetical to my main argument. It had been my plan to avoid, as I could just possibly have avoided, the obvious question about the dropping of the Atom Bomb; namely, was it a good thing to do? It is the kind of question that is canvassed in long newspaper correspondences; that leads to pulpit utterances on either side; that increases the weariness of modern railway travel by inflicting on us a good deal of "What I say is" from our fellow-passengers. In short, it is the kind of question which ceases to be unduly controversial only when it has begun to become tedious. My thesis is that the use of the bomb is in danger of giving the men and women of the coming age a guilty conscience. And it is possible to have a guilty conscience about an act which, viewed in itself, is not bad; that is the whole tragedy of the scrupulous. The best documentation of that point I know in literature is Maurice Baring's novel, *A Coat Without Seam*. The hero

RONALD KNOX

(its readers will remember) drifts away from his religion and generally makes a mess of his life because he thinks he is morally responsible for the death of his sister (from a chill caught when she was bathing), and theological assurance to the contrary, unsympathetically thrown at him, does but deepen the injury. It would be enough, then, to show that the use of this new military weapon, be it right or wrong, will in fact make some of us feel mean, and serve to confuse the moral issue for us.

But I did not want it to be thought that I was running away from an argument. Probably it *will* be thought that I am running away from an argument, because I mean to treat it on a higher level, or, if you will, on a more interior level, than is usual with topics of newspaper controversy. I have borrowed a distinction which has crept into the Oxford schools since the day when I was professionally interested in them; and I mean to discuss, not exactly whether the obliteration of Hiroshima was right or wrong, but whether it was a good or a bad thing. And there is a difference. Not that I mean to import still more philosophy into a book already somewhat overloaded with it. The difference can be stated quite clearly in terms of common Christianity, or indeed of common human experience.

Theologically speaking, my thesis is that it would have been a more perfect thing not to bomb Hiroshima. Or, if I must needs talk the language of common life, let me dig up a phrase from an almost forgotten, but not altogether unregretted past, and say that bombing Hiroshima was not cricket.

War is an anomaly. Look at it from the outside, as a neutral, and you see nothing but two parties in a quarrel who have lost their tempers at last and have taken refuge in the barbaric alternative to a just settlement, the *ultima ratio*. Look at it from the inside, as one of the combatants, and (in 1899 as in 1939) you see it as a Holy War, a Jehad, which you must needs undertake because there is no other way of bringing your villainous opponent to his senses. Having this good conceit of yourself, you instinctively fall back on a public attitude of scrupulous chivalry; because you are fighting for a holy end, you will adopt none but hallowed means; you will not soil your hands by using any form of attack which involves unnecessary suffering, which inflicts death or wounds (for example) on the non-combatant as well as the combatant. You will not sink a ship which may have women and children aboard, nay, which has merchant seamen aboard, without providing alternative accommodation for the women and children, nay, for the merchant seamen themselves. You will not rain down attack from the air upon any target except forts occupied exclusively by soldiers, or (on second thoughts) factories used exclusively for producing munitions of war. If you starve a town into surrender, you will allow

RONALD KNOX

the aged and the infirm to pass through your lines under safe-conduct. The persons of ambassadors shall be sacred, prisoners shall be treated as honoured guests. You will not poison wells from which populous centres derive their water-supply; you will not let loose noxious gases, which will take their toll of human life indiscriminately. In a hundred ways like this, you will prove to the world, and to yourself, that you are not the callous brutes which belligerency seems to make of you, which belligerency has made, alas, of your opponents.

All this, from the cynic's point of view, looks insincere enough. Modern war, he says, is total war; women, in uniform or out of it, make a vital contribution to the war effort, and every woman disabled is an asset to the opposite cause; children are the Epigoni who, twenty years hence, will be taking up arms to revenge their fallen fathers, and if they are rickety in youth, so much the better (in the long view) for the men who will then have to fight them. The plain fact is, that to drive a bayonet through your fellow-man, however plainly he is tricked out with the sacrificial garb of uniform, is a cruel business, not reminiscent in any way of the Sermon on the Mount. Of two things one; either admit that all use of violence is wrong, be patient under aggression, and leave it to posterity to vindicate you, or confess frankly that war is a kind of all-in wrestling match, and you will adopt every means to win it which opportunity puts in your way. Alternatively, if you like to draw some shadowy distinction, in the modern style, between the aggressor and the victim of aggression, admit that (in the proper theological sense of the phrase) it is the end, and the end alone, that justifies the means. Every bullet the aggressor fires at an enemy, combatant or non-combatant, means an act of bestial injustice. Whereas the victim of aggression, having testified in the sight of heaven that he is innocent of having started all this butchery, is free to drop loads of poisoned sweets from his bombers, or to put prisoners of war in a lethal chamber because there is not enough food to go round.

The cynic, it is hardly to be doubted, has logic on his side. As a long war runs its course, he observes maliciously that either belligerent has his own code of chivalry, describes as "atrocities" exactly those military expedients which suit his enemy's book, and not his own. Whichever side has the superiority in the matter of surface vessels regards the submarine as an engine of piracy; whichever side has (at the moment) inferiority in the air is loud in condemnation of indiscriminate bombing. Nor does the cynic fail to find support for his argument in the study of that intellectual offensive which we call "propaganda." It would be absurd to pretend that some of the things our newspapers wrote about the rights and wrongs of aerial warfare in 1940 were consistent with the things the same newspapers wrote in 1944. We are human; like Mark Twain's Tahitian, who, upon hearing the story of Cain and Abel, and being told that Cain was a Southsea Islander, asked the missionary, "What was Abel fooling around there for, any way?"

Not uncommonly, the propagandist seeks to avoid these embarrassments by telling us that his own side only had recourse to brutality by way of reprisals for brutality on the part of the enemy. We should do well, I think, to avoid this line of self-defence. You may defend severity in such cases, but not brutality. If a given expedient—say, that of poisoning a water-supply—is in itself wrong, then the plea that the other man did it first is no plea at all. In the heat of the belligerent emotion, it is easy to get up a popular cry that we ought to treat the enemy as they treat us. But in fact two blacks do not make a white, and we must not, St. Paul insists, "be overcome by evil." I conceive that if the enemy begin shooting prisoners of war you may rightly threaten to do the same; though even so you are letting down the standards of civilization. But if you carry your threat into effect, you are doing precisely what the Germans did when they shot civilian hostages as reprisals for the assassination of their own jacks-in-office. To shoot, in cold blood, the unarmed man who is not himself guilty of any offence is not simply killing, it is murder; you cannot claim the privilege of belligerency, for war itself has no such usage. And war must have usages to justify it, if it is to be justified at all.

When I say this, I know I am laying myself open to de-
rision from those realists who tell you that taking a man's
life against his will is an unkind thing to do, and therefore
always wrong; but when an aggressor starts a war—well, we
just do it. There are no rules; poison the wells by all means.
But these people are only preaching the immoral doctrine
I have just been taking exception to, and preaching it on
a wider scale. They want us to believe that if a person is
wicked, we are at liberty to do what we like, right or wrong,
in dealing with him; the two blacks make a white. For the
sentimentalist, this topsy-turvy doctrine will always have
an appeal, But if you are to give any intelligible account of
the matter, you must define your terms far more closely.
You must say that it is always wrong to kill a man unless (i)
you are doing it as the only way of countering an attack on
someone's person, or (ii) the man has been condemned to
death by a competent tribunal, or (iii) you and he are both
combatants, on opposite sides, in war. In such cases, killing is no murder. (I leave out of sight here the position of
the soldier who willingly supports, in war, what he knows
to be the wrong side; he did wrong, of course, in taking up
arms at all.)

I am conscious of dogmatizing; but a dogma is not a
principle you invent for yourself, it is a principle on which
a great number of people are already agreed. And here
you find agreement among the great majority of jurists,

among the great majority of Christian people, taken over an impressive area of centuries. It is the doctrine men have hammered out for themselves when they were not blinded by momentary influences—the war fever from which we suffer during hostilities, the peace fever which is apt to succeed for a time, when hostilities are over. And though there are minorities in either direction who would like us to live by some different rule, the generality of mankind will not readily listen to them until they have worked out some rival system which is consistent, and at the same time practicable.

Now, it is idle to suppose that the lives of civilians will not be exposed to more danger in war than in peace. You cannot even start shooting arrows about without the risk that you will miss your intended target and hit an innocent civilian by mistake. The risk was obviously intensified when we began to use explosives, intensified anew when we began to drop explosives from the air. But still, in theory, the civilians who are killed by enemy action are the victims not of murder but of manslaughter; it was bad luck that they should happen to be in the way, just as it is bad luck for somebody to be in the way when a car skids. The intention of the bombing was not to kill civilians; so our newspapers loudly protest, when they are on their good behaviour, and we, when we are in a judicious mood, echo the sentiment. But always the uncomfortable question is tugging at our minds, At what point does the risk to civilian life become so grave that military advantage is no longer an excuse for launching a given form of attack against a given target?

One side of this question, which was never considered at the time, has forced itself upon us in retrospect. How far is a nation at war justified in embarrassing the enemy by mere sabotage; by destroying means of communication, mines, factory buildings, crops and so on, *with the certainty of producing economic distress when the war is over*? Distress, we may add, not only to its enemies, but to harmless neutrals, nay, to its own allies who are the victims of enemy occupation? In a word, are we even well advised, let alone morally justified, in giving *carte blanche* for destruction to a set of air chiefs whose only aim is to do as much damage as possible, whose plans are so shrouded in secrecy that all outside criticism of them is impossible except on what is nowadays called "the highest level"? The familiar irony of Southey, "Things like that, you know, must be, after a famous victory," if it was pointed in his own day, is triple-barbed in 1945. In a few months' time there will be peace-propagandists telling us that Southey was right, that war can be avoided; once it breaks out, its horrible indirect effects are inevitable. But are they inevitable? If so, God help us.

In the years between 1918 and 1939 we overreached ourselves by our own idealism. We were so determined to banish war from the world that we refused to discuss, in any effective way, the proposals that were made for humanizing the methods of it. To have a convention against air attack upon undefended towns seemed almost blasphemous; it was like licensing brothels—you were parleying with the forbidden thing. Are we going to adopt the same policy again? Much propaganda has been made out of the atom on those familiar lines. This new engine of destruction, we are told, will make war *impossible* in future. Yet we were told the same, in the peace-propaganda of the twenties, about the influence of poisonous gas on the future of warfare. And while we told one another, with wise shakings of the head, that poisonous gases had made war impossible in future, we were paving the way for the bloodiest war in all history, in the course of which neither side dropped a gas bomb.

At the moment, we are in a chastened mood, and the prospect of a warless world seems uncertain; we are for regulating the use of the Atom Bomb. When some convention meets to discuss it, only two views of it are effectively on the agenda. One is, that the use of such a weapon is in itself wrong, and must therefore be abandoned universally. The other is, that it is a very terrible, though a tolerable weapon, and it must only be used in war against an aggressor. The former decision will mean that we did something wrong in 1945. The latter decision will mean—since no nation regards itself as an aggressor nation—that mutual fear, rather than mutual good-will, is our best guarantee against a world left uninhabitable. If a year or two is allowed to elapse before the convention meets, I think we shall be sorry, by that time, that we used the Atom Bomb.

As I have said above, I do not mean to discuss here whether its use was right or wrong: nor even to discuss whether the bombardment of populous cities, which happen to be important railway junctions, is right or wrong. I only say that it would have been a *good* thing if we had left the new weapon unused. It would have been a gesture of generosity, which might have had incalculable effects on the future of war; which would have given some hope of the Pacific living up to its name; which would have offered an inspiring example, as I hope to show in my next chapter, to the conscience of the individual citizen. And it is too late.

Unless you take a narrowly Puritanical view about works of super-erogation, you have to admit that *a given* course of action may be *right*, yet a different course of action may be *better*. King David, when his three heroes broke through the enemy's lines and brought him a cup of water from the well of Bethlehem, poured it out as a libation instead of drinking it. The sacred author implies that he did a better thing. On another occasion—on two other occasions, apparently—he had the life of his enemy King

Saul in his hands, and spared him. It is difficult to believe that the conscience of that age would have condemned him if he had satisfied his desire for vengeance. But we shall all agree that he did a better thing. And, granted that we had a right to destroy Hiroshima, I say we would have done a better thing if we had emptied that fatal cargo, in the first instance at any rate, by way of demonstration, upon some untenanted mountainside. Like David, holding up the piece of cloth from Saul's garment in the faint light of dawn, we should have done a better thing by showing the Japanese what we might have done, and not doing it.

Suppose a candidate, let us say at a Parliamentary election, whose opponent is an old hand at politics, at once popular and unscrupulous. He is warned by his agent that, if he is to win the seat against such competition, he must leave no stone unturned. And to win the seat is, for him, not merely a means of gratifying private ambition; he honestly believes that the programme for which his rival stands will be disastrous to the country, and, in his efforts to convince the constituency of this, he is quite ready to hit out, and to hit hard. And now, by sheer accident—an old letter left about by mistake between the pages of a book, the gossip of a servant, whatever you will—he comes across convincing evidence of a fact calculated to bring odium on his rival. Some fact quite unconnected with politics, and damaging only in a general way; the man has treated an old servant of his shabbily, his jockey has pulled a horse, or what not. The fact could not be publicly mentioned without fear of libel proceedings; but a word to the agent would mean that the story would get out, it would go the round of the public-houses and the hairdressers; silently but effectually, it would influence public opinion. Shall he give the word, or not?

If the word goes round, there will be no grave consequences to the rival candidate; no pecuniary loss, no legal proceedings, no loss of good name, even, in his own circle of friends, for he is not a local man. There is no question of a breach of confidence; the secret was betrayed by mere accident. Difficult, therefore, to claim that the man who has hit upon the discovery has not a right to make use of it. And, it may be urged, if he has a right, surely he has a duty to make use of it? At all costs, the undesirable candidate must be kept out, in the public interest. And yet, if the man who has the secret in his possession is a man of nice conscience, he may, I think, quite probably, and quite properly, keep the secret to himself. His private discovery is his very own, to do what he likes with. He has a right to divulge it, but he has a right not to. And he exercises that right by indulging in a gesture of generosity.

I have been assuming that the discoverer of the secret is one who thinks in theological terms. You do not necessarily have to think in theological terms before you can exercise restraint of this kind. I have the picture in my mind of a large, leisurely Englishman of yesterday taking precisely the same line in the same given circumstances not (consciously at least) from any supernatural motive, but because restraint of this kind is part of a code. Without stopping to ask whether it would be right or wrong to divulge the secret, he dismisses the idea at once from his mind as "not cricket." I am not concerned here to defend this rather demoded figure against the strictures of the theologian, who will tell me that this is merely "natural morality," or against the strictures of the modern progressive, who will tell me that there is little virtue in such a code; an aristocratic class, which has long accumulated the rewards of unjust stewardship, finds it easy to practise a justice exceeding that of the scribes and Pharisees. I am only concerned to point out that you do find, even among natural moralists, the instinct of making a generous gesture, of going one better than was to be expected of you. On the value of their motives, let a higher Court decide.

Such gestures, of course, are far more easily made by a private individual, whose own interests are mainly concerned, than by a set of statesmen charged with the destinies of a great alliance. Other men's lives are at stake; those, for example, of British or American airmen who might be shot down in trying to pin-prick the targets of Hiroshima one by one, instead of devoting it to a general holocaust.

But I still wish that the gesture had been made; that the Western Allies had said, in effect, "Here is a new weapon; see, from the effects it has produced on yonder deserted camp, what would have happened if it had fallen on one of your cities. The Germans were on the track of it; had they discovered it in time, we should have had no warning. It has been put into our hands instead; it shall lie safe there. We have enough civilian blood on our hands already; Russia is in the field, and the war cannot last long now. Let this libation be poured to the future of humanity."

Would the moral have been lost on our enemies? Not, I think, in the long run. They might have thought us fools, but it would have impressed them to see that we had a *bushido* of our own. Meanwhile, it would have restored, incalculably, our own self-respect. To fight with the gloves off may be an invigorating experience while the actual crisis lasts; the mysterious instinct of *schadenfreude* peeps out in all of us. But when it is over, and you look round at the wounds that have been inflicted on the world—not only the outward world, but the world of men and women—a different mood succeeds. A mood, not of repentance exactly, but of that ruth which is cousin to repentance; you tell Little Peterkin to stop rolling that round thing about; hang it all, there are limits. This chastened mood we seek to compensate in unreasonable ways. In our public life, by an itch for changing everything; by rash experiments,

and undignified recriminations. In our private lives, by
reminding ourselves, that after all, the Ten Command-
ments seemed to go west when there was a war on; isn't
it rather hypocritical to go back to them as if nothing had
happened? Is there any sense in pretending to ourselves we
are better than we really are? We are ready to let the yoke
of duty drop altogether from shoulders that have borne it,
yesterday and the day before, so unworthily.

꙳

VI

Control and Release

GOD FORBID that relations between man and man should be modelled on those of rival peoples when they are at war. We have all come to accept it as a regrettable fact that policy in war-time is only a by-product of strategy; and we grant the diplomatist the same latitude which is claimed by the general in the field of concealing, and even misrepresenting, his designs. We should not care to do business with a man who had learned his sense of truthfulness in a Ministry of Propaganda. And there is in diplomacy, apart from the question of truth, an undertone of bullying, blackmailing and bargaining which might pass muster in big business, but would scarcely endear a man to his friends. Our lives depend increasingly on Whitehall, but we do not go to Whitehall to learn how the good life should be lived.

Yet there are repercussions; the citizen is not content in the last resort to admit that there is one standard of morality for his country, another for himself. We still hanker after the Platonic notion that you must look for justice in the State, where its pattern should be written large, before you learn to recognize it in the life of the individual. That is why I devoted my last chapter to considering the possible effects of a gesture which was never made. If the Allied Powers had shown us an example of self-restraint, when they were tempted to go all out for their own interests, it would have been an encouragement to us struggling mortals, who find ourselves, from time to time, in the same position. As it is, when you or I see the opportunity of winning some coveted prize by the use of unscrupulous means, there is always danger in the reflection, "What did the Allied Powers do when they found a new weapon at their disposal?"

Self-restraint is a quality the Atomic Age will not find it easy to come by. The regimentation of life from above—controlled prices, rationing, compulsory national service, direction into trades, taxation of excess profits, and so on—does not, of itself, evoke a spirit of discipline in the citizen's mind; on the contrary, the more regulations hedge him in, the more resolved is he to evade them, the more fertile does he become in expedients for their evasion. We were warned of that, when Prohibition was introduced into the United States of America; whatever good or harm that experiment did, one thing is certain, that it created a new and very flourishing class of lawbreakers. Rum-running was the satellite of Prohibition, as smuggling of the excise duties. And in the last few years we have seen the birth and growth of a Black Market, whose promoters are not merely rebels against a human law, but in many cases (though no doubt they do not ask questions) receivers of stolen goods. Nor does the ordinary citizen remain unaffected. At moments of crisis, during the air-raids, for instance, he has not ceased to be the kindly Englishman we knew, always ready to take trouble over people who appealed to him in distress. But long hours of waiting in queues and standing in trains have developed in him a habit of self-assertion, of letting the devil take the hindmost, which you can see increasing from one year's end to the next. The man who gets up to offer his seat in a railway carriage to a woman—a plain woman, anyhow—is almost as noticeable as the man who takes off his hat when he goes into his bank. And it is certain that during the next few years we shall still be galled by the shackles of Government control, without the enthusiasm of a nation at war to make discipline more tolerable.

Self-assertion, an instinct which obviously lies at the root of us, civilized creatures though we are, will come to the fore, and we shall need strong checks within us to counteract it. Now, it is regrettably true that one of the strongest checks which operate to deter a man from crime, or from disgraceful conduct, is the want of practice. Supposing that a school, as part of its regular curriculum, should train all its pupils for an hour a day in the art of counterfeiting signatures. In itself, it is an amiable accomplishment, and no doubt it might be claimed that a boy who devoted himself to this art would develop his powers of observation, steadiness of hand, and other useful qualities. But we should all have an obvious objection—that the average employer would prefer to take on a clerk who had no skill in this art of counterfeiting. It is no use for the headmaster to assure us that he is not *teaching the boys to forge*; the fact remains that he is *teaching them how to forge*. And in teaching them how to forge, he is breaking down one of the safeguards of their innocence. To be sure, in strict morals the fully formed intention of committing a crime has all the guilt of a crime, even if you fail to bring it off; and a clerk who is *only* deterred from forgery by the dread of being found out is in no very creditable moral dispositions. But facility does increase temptation, and there is little doubt that the headmaster of our parable would be invited to send in his resignation.

War is a school of violence. Thucydides noticed, long ago, how the long years of the Peloponnesian War brought with them a deterioration of public morals; crimes of violence, he assures us, became more frequent, simply because the conditions of the time had made acts of violence

more familiar. Even in England, we have been training
men to butchery, and to butchery by stealthy means, as
an activity valuable to us in a life-and-death struggle. We
have selected, not a set of criminals released from jail, but
some of the best, some of the most promising young men
we could find, and taught them to be good assassins. I am
not criticizing the policy which dictated this step; still less,
Heaven help us, am I criticizing the men in question—we
must needs reverence them as heroes. But the fact remains that there are probably ten times as many men in
England who could creep up noiselessly behind an unconscious victim and kill him with a single blow as there
were in 1939. All over the continent of Europe the situation is even worse. Young men have passed their apprentice years, most of them from purely unselfish motives, in
secret and relentless revolt against the *de facto* government
of the countries they belonged to. They have adopted an
honourable career of brigandage; and already we are becoming conscious that they do not always settle down into
harmless, law-abiding citizens, merely because an official
intimation has been issued to them that a state of war no
longer exists. There are still private vendettas to be worked
off which make them reluctant to hand in their weapons
to the official police. Their consciences are quite unaccustomed to draw any distinction between war and assassination. Will it be easy, in a world very short of supplies, to

RONALD KNOX

imbue such minds with the principles of John Stuart Mill?

It may be objected that we have strayed far from the atom; I am not so sure of it. Once more, I would entreat the readers of this book, and even the reviewers of this book, to realize that I am not concerned so much with the conscious motives which operate in men's minds, with their reasoned calculations, as with their mental background; with the images, the associations that are called up to their minds when need urges, or opportunity offers. Is there not reason to fear that habits of self-restraint will come to our children all the less readily, simply because the newspapers have encouraged them to believe that, in some mystical sense, they are citizens of an Atomic Age? To some extent, our minds are always conditioned by the world-picture that is fashionable in our time. To some extent, this conditioning is responsible for our unreflecting actions. And to that, extent (it seems to me) the Atom will be a dangerous counsellor.

So long as our minds lay under the spell of the Machine Age, the danger was deadness, was unimaginativeness. We might be tempted, indeed, to abandon moral effort, in despair of overcoming our own weaknesses; we were "made that way," destined to run along a particular groove, and it was the part of wisdom to accept our destiny. But at least the picture we had formed of the external world encouraged us to run well in harness. In France, an

So long

ignorant peasantry might turn against its rulers; we, thank Heaven, knew our place better than that. There should be nothing exorbitant about our conduct; we were born to the rut. The pathetic protest of the Luddites, breaking up the looms for fear that machinery should displace men, was a parable of all our efforts to revolt against our surroundings. *Men* had already been displaced; by machine-men, described for the first time in the eighteenth century as "hands." Self-assertion could only be a kind of sabotage, while we thought of ourselves as human cogs in a universe which was all cogs.

The evolutionary world-picture was plainly more encouraging to private effort. If we were to adopt the Stoic maxim, and live according to nature, we must adapt ourselves to a struggle, and be Nature's favoured children in a fiercely competitive world. If not, we should go under; the age of *laissez-faire* was an age of *laissez-mourir*. But even now, though we were bidden to bestir ourselves, there was no incitement to kick over the traces. Evolution gave you the image of gradual, not sudden, rise and decay; and it was as part of a vast world-system that each species decayed or rose; you had to fit into a background. You were not content, to be sure, with the economic notions suggested by the Industrious and the Idle Apprentice. In a world which no longer believed in an eternal delimitation of natural species, you were not going to believe in an eternal delimitation of social classes. But if a man aspired to rise out of his class, he must do it by conforming to the laws of the social hierarchy. And even if his ambition was more unselfish, to rescue a disinherited class from its depressed condition, he must still think, not in terms of violent catastrophe, but in terms of slow upward development; abuses must die out gradually, like the dodo; the working man, like man himself, must come to his own in orderly fashion, not by fits and starts. For the treadmill ethics of the Newtonian Age, the Darwinian Age substituted a doctrine of progress; but it was progress of an orderly kind, slowly broadening down from precedent to precedent; all would be well in the long run, but in Heaven's name let us keep step.

That was still, I think, the world-picture of the Edwardians. You did honour to the self-made man, but you expected him to wear a clean collar. You accepted, either with or without enthusiasm, the claim of the manual workers to be better paid, to have more freedom of the world's amenities, to have their children better educated; but it was to happen gradually, *that* went without saying; we should see the income-tax slowly broadening up from n to n plus one percent, and Utopia would be among us before we knew where we were. It is only during the last twenty years that we have come to accept catastrophic change in the government of neighbouring peoples as a normal feature of the day's news. And, to the older among us at any rate, with our Darwin-conditioned minds, every such incident has come as a mild shock; not thus (we felt) the amoeba rose to greatness.... In the nick of time, it would seem, the new notion of atomic force has come in to supply our juniors with a suitable image for their day-dreams.

Dimly as the lay mind comprehends such mysteries, the gossip of the laboratory does give us some idea of what this force is like; and it is like nothing we knew to be on earth. For one thing, you must set about destroying the atom if you are to release the power latent in it; is destruction, then, after all creative? The effects of your action mount up, it appears, in geometrical, not in arithmetical progression; is this the prelude, we ask, to a world in which all the news is stop-press news? Atomic power has been manifested to us, so far, only as an instrument of death; and the bomb (like all explosive weapons, but on a scale hitherto unimaginable) is a weapon in the hands of tyranny. It is suited to the needs of a world in which you no longer count heads to save breaking them, but blow off heads to save the trouble of counting them. Democracy labours for breath, when the power of mass-murder is concentrated, for good or evil, in the hands of a few. And our prophets tell us that when atomic power is harnessed to peaceful uses, it will displace human labour by machinery on a grand scale. In proportion as that happens, democracy is robbed of the only weapon it has invented as a substitute for revolution—the strike. In 1918, we talked of a world made safe for democracy; the world of 1945 seems a world that is safe for nobody, but with a slight margin in favour of the gangster.

But I am not concerned with the politics of the Atomic Age, as such; I only mention them as illustrating, and perhaps helping to form, the outlook of the individual citizen who is reared on such milk. It is reasonable, I think, to suppose that his cast of mind will be subtly different from that proper to the Newtonian Age, from that proper to the Darwinian Age; that he will disagree with his elders not in a simple, evolutionary fashion, like Stevenson's tadpole saying to the frog, "Just what I thought, you never were a tadpole," but with a fundamental difference of bias which threatens to put us all at cross-purposes. Youth loves to revolt from the maxims of its fathers, but ordinarily with the sense that it is only following out the implication of its fathers' logic. "You bid us follow nature," the Victorian world said (in effect) to the Georgian, "by living in a groove. But you have not realized that Nature's groove is an upward one, a groove of change; if we would be a part of nature, we must take part in her struggle." But the new age will surely be at issue, and fundamentally, with both. Instead of taking matter for granted, like the eighteenth century, it will by-pass matter and think in terms of power. Instead of concentrating attention, like the nineteenth century, on the idea of slow, automatic progress, it will be inspired by the notion of incalculable forces suddenly released. Its emblem will be, not the bud that bursts into a flower, but the spark that bursts into a flame.

I suspect, then, that those minds which are now in the making will be a bad surface for receiving the polish of altruism. The appeal of religion will come to them as a thing alien from their own unconscious prepossessions. The days are long past when philosophy was described as the handmaid of theology; the most we can expect nowadays is a little occasional help. And science, equally, has played us false, if we had any hopes in that quarter. For Nature, more than ever, seems to have been reduced by science to the position of a drudge and an accomplice; she lends herself to our purposes, and such purposes! Yesterday, we pointed to her with reverent fingers, "See, how her mysteries still baffle us! See, what a lesson we can derive from the majestic leisureliness with which she works out her designs!" Deified yesterday, today she is defied; we are close to the heart of her secrets, and we know, now that she was cheating us; that all her boasted orderliness was but a screen for anarchy. Will they not tell us, the men of the new generation, "We are atom-children; do not be surprised if we turn out worthy of our breed?"

RONALD KNOX

Yet never was self-restraint more necessary, if we are to have a habitable world. The whole economic machinery of Europe has suffered a break-down; the business of repairing it will be slow, and it will have to be run in gently. There will be controls everywhere, and all the prizes of life will have to be queued up for, unless we want a free fight. There will, please God, be enough to go round, but only just enough. And comfort means, not having all you need, but having a little more than you need, so as to leave a margin. To be always counting the biscuits and measuring the remains of the pressed beef, to work and sleep in the same room, to pinch yourself when you are going to entertain a visitor, save up for a fortnight so as to make a spread for some festive occasion—all these are not hardships, but they interfere with the smooth running of life; it is like a hard chair without a cushion, a tap without a washer. There must be a certain largeness in our surroundings, if we are to be effortlessly happy. To make your career, to marry a wife, to have a family, in a world where everything is a tight fit, robs youth of its pleasant carelessness, makes it more apt to brood, more selfish, more ready to revolt against its chains, where revolt seems possible.

All the more so in a society which accepts it as an agreed maxim that "the children must have the best of everything." All our sentiment, all the wise warnings of the psychologist, plead for such an arrangement, and he would be a bold man who should make a bid for reversing it. But it has the disconcerting consequence that emerging from childhood into your later teens is no longer an emancipation, the entry into a larger world, as it was a generation ago. On the contrary, it is the adolescent that begins to feel himself or herself unwanted; school days, instead of being a thraldom, are a sunshiny world with a dark tunnel at the end of it. I do not pretend to trace, with any certainty, the influence of this chilling consciousness on young lives. Bad enough for the layman to philosophize, to grapple with the mysteries of science; it is still more difficult for anyone within sight of sixty to recapture the sense of what it feels like to be young. But it does seem obvious that in these next years the rising generation will be called upon to exercise a self-restraint uncongenial at once to its time of life, and to the instinct of its period.

And the word "religion," twist it as you will, means restraint. By its derivation, it means that dread of supernatural consequences which keeps a man true to his sworn oath when there is advantage to be gained, in this world, by breaking it. Of its nature, it means self-dedication; signing away your independence, to give priority to Something which, *ex hypothesi*, claims absolute priority—a fiction, say the moderns. And this Something underwrites all the dictates of conscience, hamstrings the mind with hesitations when there is a blow to be struck for self-interest;

reinforces the claim of altruism, and sends us back to the tail of the queue. Oh, I know this is a miserably inadequate account of religion, taken in its whole extent. But it is the guise under which religion presents itself to the ardent spirit already dimmed by a sense of frustration. Especially when religion, to suit the exigencies of an agreed syllabus, has been handed down chiefly in the form of a moral code. For a scruple, you are asked to let the unrecoverable opportunity slip through your hands, let the other man get in first. To have the world at your feet, and say No....

That is why, I repeat, I wish the Allied Powers, with the world at their feet and the Atom Bomb in their hands, had said No. Do what we will, war on the grand scale leaves us with a nasty taste in the mouth. We do not quite like to admit that we were wrong to engage in it; that would be to justify the claims of our enemies. But all the more because we cannot very well repent of it, we are subtly ashamed of it; and this sense of shame has to be rationalized, usually, by attacking our leaders. Either we tell our political leaders that they went the wrong way about it, used the wrong weapons, made the wrong alliances. Or else we turn upon our religious leaders, and ask why they were so ready to pronounce their benediction on an orgy of mass-murder—forgetting how, at the time, we pelted with abuse those other religious leaders who tried to restrain us. The Churches, we say, have made themselves look ridiculous

and worse by lending themselves to the service of chau-
vinist propaganda. Organized religion is the scapegoat we
drive away into the wilderness, loaded with the burden of
our own unspoken remorse. But, deep in our hearts, the
scruple still rankles. There is blood on our hands, and not
all the perfumes of Arabia will wash them clean.

I cited, at the beginning of my last chapter, a novel of
Maurice Baring's in evidence that a scruple undislodged,
no less than a sin unrepented, can give a man a guilty conscience and leave him a victim to the moral laxity which
is born of despair. So it can be, I think, with a whole nation; so it was with us, a matter of ten years after the first
European War ended. Not admitting for a moment that
we had been wrong to go to war, we managed to convince
ourselves that it was always wrong for anybody to go to
war. And this scruple robbed us of our old confidence that
Great Britain had a glorious record of generous altruism
which we, as good citizens of it, must needs live up to. We
tacitly acknowledged in ourselves a kind of moral sec-
ond-rateness which served as an excuse for low standards;
we were poor creatures, and morality must not expect too
much of us. I may be wrong, but I anticipate a similar reaction not many years from now, which will threaten to
plunge us still lower into the depths of self-abasement, and
of consequent self-despair. In what precise way we shall
rationalize this mood of ours, *I* do not attempt to predict.

But I think what will chiefly help to fasten it on us is the memory of having called in a sinister weapon to win the fight for us, a weapon which may recoil on humanity, and possibly on ourselves. We shall feel like men that have sold themselves to the devil; we have conjured up the Atom, and the Atom henceforth is to be our master.

St. Augustine says that nobody serves the devil willingly. Well, no, perhaps not if you put it like that. But there is a certain glamour about the leadership of a splendid rebel, and I suspect that the Atom, by then, will have found its conscious devotees. No doubt, by then, it will be doing all sorts of useful things for us, and we shall be able to think of it as a sort of Robin Goodfellow, that has come in to do drudge's work, leaving its dangerous possibilities out of sight. But though it may enable us to close down our mines, to outdistance all our speed records for land, sea, and air, to make diamonds out of coal, and to establish a twenty-hours' working week, all *that* may be purchased at too dear a price; we shall be the slaves of the Atom, not its masters, if we consent to wear its livery—the livery of explosion and of revolt.

I take it that we do not exceed the bounds of legitimate metaphor, if we think of the human personality in this way. At the core of it, there is a bundle of instincts, impulses, prejudices, phobias and what not, each of them bound, and each, though often in a very slight degree, straining at its bonds. They are held together and held in by the elastic band of Repression; some of it conscious, much more of it unconscious, or half-conscious at the best. If the band snaps, the result is lunacy; all the hidden impulses of a man's nature regain their freedom, held in only by random, external checks. If the band slips, the result is that sudden brain-storm or black-out which the psychologists have christened schizophrenia; the subject "forgets himself," is untrue to his normal habits of behaviour; it may be, only for a short interval. But in the ordinary life, the elastic band holds, and the hidden impulses remain bound, only betraying themselves by casual mannerisms and fidgetings, by the images that haunt us in our dreams, and so on. What must be the strength, when you come to think about it, of this band which holds our psychic life in position, consisting in part, but only in part, of that free will which we consciously exercise!

Now, in the atom, the unsplit atom, there is an energy (Hiroshima knows how terrible) which is engaged, and from a certain point of view is going to waste, in holding the atom together. It is the liberation of it that is the characteristic triumph of the new age. Will there not be a temptation for some child of the new age to interpret his own personality as follows? "Here am I, exercising a prodigious force, unconsciously for the most part, in the dull, tame business of keeping my own impulses in check.

RONALD KNOX

A prodigious force of will, going to waste (you may say) for all the good it does me, serving the purposes of a set of taboos, religious, political, social, conventional, constraining and falsifying my true nature. Such a misapplication of force is surely unworthy of the Atomic Age. I must set myself to liberate this force of will, so as to fulfil a positive, not merely a negative function. Instead of hampering me in the achievement of my ends, it must be so canalized as to help me in the achievement of my ends. It must be trained to form the motive power, not of self-repression, but of self-assertion." Once more, let us remind ourselves that this is not likely to be a conscious calculation. In our conscious calculations, we are too much on our guard against the danger of losing our reason to argue like that. My fear is rather that in the background of their minds men will be arguing like that, and flattering themselves that they are making economic use, at last, of their hidden reserve of will-power, when in fact they are only unchaining this impulse or that, for their own self-preservation, their own self-aggrandisement.

Thus, then, I would complete the picture of the disturbance which may have been made, in the mind of our generation, by the trauma of Hiroshima. It is not only that we shall feel vaguely as if our certainty about the existence of a God no longer applied—the new physics have fastened on us a method of thought, a recipe for organizing our experience, to which a spiritual Co-efficient is no longer necessary. It is not only that we shall feel vaguely as if the world were no longer Providentially governed—if indeed it ever was! As if Man had taken his destinies into his own hands, who knows with what appalling results? Besides, all this we shall feel vaguely, some of us at least, that the atom is the symbol of our release from every internal principle of self-control. "So many checks and regulations fettering us, in these days, from without—the mind, at least, shall be a kingdom of its own, or rather, an anarchy of its own! World-domination, after all, belongs not to the cause which has right on its side, but to the cause which has the best equipped laboratories on its side. We, too, will be fiery particles, bombinating in a world of unrest."

ADJUSTMENT

VII

An Alternative to Doubt

MY THIRD chapter lies open to one obvious criticism. I have written as if belief in God depended entirely on those five scholastic proofs which take as their starting-point, not the knowledge we have of ourselves, but the knowledge we have of external nature. To argue by inference from effect to Cause, from the passive object to the active Subject of change, from transitory, contingent being to a Being who is necessary and eternal, from nature's striving after perfection to a Perfection which is ultimate, from the order observable in creation to a creative Mind—all that (I shall be told) is to approach the great Riddle from one side, and that the most difficult. Meanwhile, as we all know, other proofs have been offered, and seem to many minds, in our day at least, more cogent.

There is the ontological proof, in its various forms. As stated by St. Anselm, it is content to ask how God, who is *ex hypothesi* perfect, could lack so important a perfection as that of existence. With Descartes and his followers, it asked rather how we are to trust any of our own mental processes, if our finite intelligences are not underwritten by an Intelligence which is infinite. Kant brought us back to the argument from conscience; we had the inner assurance of being at issue with the dictates of a Will, surely not less personal than our own. For the Hegelians, there must be an Absolute, to transcend the complete gulf that lies between subject and object in our experience. Today, the argument from the "numinous" is in fashion; the very fact that we have an instinct of worship, that we feel a sense of awe (rightly or wrongly) about such and such a place, or name, or department of life, is our best guarantee that a supernatural world exists; this sense of awe is not to be confused with any other sensation.

I should be the last to find fault with these other methods of approach. To many, they will appeal as more direct, more intimate, perhaps more profound. But they are, it must be confessed, the afterthoughts of the introvert. For the generality of men, the world of their outer experience is the real world. I remember, long ago, the late Archbishop of Canterbury describing to me his attempts to argue a working man out of his materialism on Hegelian principles; all he got was, "Ow, don't talk like that; you make me feel quite funny." What is most familiar to us is not what is nearest to us, but what we can hold at arm's length. And,

although the five classical proofs may seem abstract and
arid in these days, when we have grown unaccustomed to
the language of metaphysics, they are nevertheless a reasoned statement of the conviction most men either hold,
or wish they could hold; namely, that things seen are the
work of an invisible Creator. That is, after all, the faith on
which their childhood's confidence was grounded. "I will
consider the heavens, the work of thy fingers.... Consider
the lilies of the field, how they grow"—it was these ele-
mentary considerations that were proposed to us when,
in the first dawn of doubt, we asked why we could not see
God up in the air. Alternative arguments, however valid,
always produce the vague suspicion that they are by-pass-
ing a difficulty. If God did not create heaven and earth,
the fact of his existence is not particularly impressive. If he
did, why could he not write his *pinxit* at the corner of the
canvas, instead of leaving the attribution of the work to be
a matter of inference?

I have called the scholastic argument a "reasoned
statement" of our conviction that things seen are the work
of an invisible Creator. It is an old complaint, that metaphysical proofs may be all very well for philosophers,
but that the faith of the charcoal-burner must depend on
some other kind of certitude. The answer is surely that
every man apprehends truth at his own intellectual level;
one will depend more on picture-thinking, on crude,

unanalysed notions, than another. No doubt but Man, at a low stage of intellectual development, does "make God in his own image." He feigns, and half believes, that the thunder is a divine shout, the lightning a divine arrow, the sunset clouds a divine dwelling-place, and so on. At that stage, if he speaks of God as "creating" the world, he feigns, and half believes, that an enormous Giant literally piled up the hills, like a child making a sand-castle, literally poured the water round them, and fixed the stars in the sky like pins in a pin-cushion. What he is struggling to express is a truth not less, but greater than this; namely, that the visible world of our experience came into existence through the will, and according to the plan, of a wholly spiritual Being. It is only by a metaphor that we describe that Being as having hands or fingers, as "dividing" the sea from the land, or even as "speaking" in accents of command. In cold fact, we have so little understanding of the influence of spirit on matter that our nearest approach to an accurate statement of it is by way of causality. We are familiar with the reflection that such and such a thing would not have happened, but for such and such an antecedent condition; the incidence of a particular disease, for example, is due to the lack of a particular vitamin. We are so certain of this principle that we often postulate a cause to account for a given effect; the Glacial Age, for example, and the melting of the ice, to account for particular effects in geology. And if we push our demand for explanations far enough back, we reach a First Cause, That without which nothing would ever have happened at all.

In a word, when the savage points at the sun and tells us that a very tall man must have hung that round thing up there, we reply, "Well, that is an anthropomorphic way of putting it, but I see what you mean. What you are saying in a rather highly-coloured way is what *we* mean when we say that every event demands a cause to account for it, in the large, no less than in detail. If this material world came to exist, if life supervened on matter, sensation on life, consciousness on sensation, all this demands a Cause to account for it; meanwhile, no cause or set of causes in the material order can be commensurate with the effect produced." Now, I am fully conscious that there are moderns to whom this way of talking will sound hardly less out of date, hardly less *naïf*, than the unaided speculations of the savage. But I want to hear these moderns put me right, just as I put the savage right; I want them to explain more accurately what it is that I am trying to say, just as *I* did with the savage. But in explaining this, I am not going to let them explain it away.

Do not let them tell me that when I talk about cause and effect I mean invariable sequence, or invariable concomitance, because that is not true. When I say that the emergence of life on the globe must have been caused by

RONALD KNOX

Divine intervention, I do not mean that each emergence of life on every globe is invariably accompanied by Divine intervention; how could I? Thunder and lightning come together in my experience, but not for a moment am I tempted to think of either as *causing* the other. If A always follows B, if I always have a sleepless night when I have mushrooms for dinner, I may begin to *suspect* that the mushrooms are responsible for the insomnia; but it is obvious that in going through this mental process I am distinguishing between the idea of plain concomitance, and the idea of cause. Do not let them tell me that my notion of invariable sequence has gradually grown up into, gradually shaded off into, my idea of cause—as (they tell us) the eye, from being an organ of touch, has gradually grown up into, gradually shaded off into, an organ of sight. For this is to confuse the form with the content of an experience. Blind men do not see ghosts; the savage may think, foolishly, that his beating of tom-toms has made the moon come out of eclipse, but only because the idea of cause and effect is *there already in his mind*. Do not let them tell me, even, that the notion I have of cause and effect is something which I derive from my own experience as a free, conscious agent. When I say, "I have been the cause of this man's misfortunes," I am expressly distinguishing between what I meant to do and what has in fact resulted from my behaviour. Finally, do not let them tell me that I have an innate idea of causality, antecedent to and irrespective of any contact with the outside world; for this is mere mythology, untrue at once to reason and to experience.

No, let them analyse my notion of cause and effect, make it more explicit, show me its bearings, relate it to the rest of my experience, shed new light on it, and I will thank them for taking pity on my stupidity; but do not let them filch it away from me by offering me counterfeit tokens in exchange. Meanwhile, the fact (if it be a fact) that scientists do not, in our day, find this particular tool of thought valuable to them in their researches, has no real bearing on the issue. If we were pragmatists, and thought truth depended on practical utility, then indeed a process of thought which had lost its value for science would have lost its value for philosophy. As it is, I am more tempted to suppose that a process of thought which does not help me to discover anything about the nature of the atom must be designed to help me in discovering something about the nature of Almighty God. But indeed, the human mind works so variously, hits upon its ideas so much at haphazard, that the practical results it achieves are no real criterion of its methods. Columbus discovered America under the impression that he was approaching the Indies from the east, but he discovered America all the same. Who cares whether Jenner thought of immunity from small-pox as the effect, or merely as the invariable concomitant, of the dairy-maid's profession? The world owes its debt to him, not as a philosopher, but as a scientist.

But the plain man, groping his way through the revelations of modern physics, has a worse puzzle in store for him. If the pioneers of research are content to by-pass the notion of causality, there is no great harm done; what is more annoying than to be blamed by your host for not taking some patent short-cut of his, on the way up from the station? It is more serious when they question our dogma that nature is governed by strict law, and want us to leave pure chance to take a hand in its destinies. *Pure* chance; not merely chance in the bookmaker's sense, which means that we (and possibly he) cannot predict with any certainty what is going to happen, but such indeterminacy as could exist, we thought, only in human wills, and perhaps not even there. It is as if we woke up one morning to find plums growing on the apple-tree, and the gardener telling us that, on the law of averages, that was only to be expected.... And if we ask why this anarchy at the heart of nature is not reflected at any higher level, why apples do grow on apple-trees, and not plums, we get the impression that the law of averages comes in, creating the illusion of a uniformity which is not really there.

It is perhaps worth observing that the proof from Order as given by St. Thomas does not insist upon a completely verifiable law of uniformity in nature. He says, moderately enough, that if we see means being adapted to ends universally *or for the most part*, we can legitimately infer the existence of a Mind responsible for the adaptation. If, therefore, the physicists are content to tell us that one class of substances—radio-active substances—form an exception to the law of uniformity, this will come as a severe shock to the mind of the eighteenth century, but not necessarily to the mind of the thirteenth. For the Age of Mechanism, the law of uniformity was a dogma; human thought stood or fell by it. If you had fully entered into the spirit of that age, you were loath to admit even the truth of the Gospels, where they recorded the miraculous; once broken, the great Law would be broken beyond repair. But the schoolmen, looking round upon the external world in a less exacting fashion, were conscious of a great deal of uniformity going on around them, and that was enough for their purposes. They could not be bothered to investigate all the tales of strange phenomena that were then current; possibly the salamander *did* live in the fire, but that did not alter the general rule that if things came in contact with fire they got burnt. The great bulk of their experience showed them the picture of means being adapted to ends without human intervention; and how could you account even for a seventy-five percent uniformity in nature without positing the existence of a regulating Mind? A few sequences here and there, which looked like cause and effect, might have been written off as mere runs of luck. But this steady uniformity about the vast majority of observed events was not to be accounted for by accident.[†]

Accordingly, if you had told St. Thomas that the radium atom exploded when it liked (so to speak) and not in accordance with any law whatever, he would not have been greatly exercised over it. But all this is to assume that radio-activity is a phenomenon quite unlike anything else in nature. And what the scientists seem to threaten us with (when they have gone still farther into the matter) is a picture of *general* anarchy at the other end of the microscope; a picture of tiny things moving about by blind chance and only, as I say, achieving uniform results because the law of averages steps in, and sees to it that by and large the same rule holds good every time. Not an Infinite Mind, but a mathematical formula, is responsible for the rhythm you and I, without microscopes, trace in the world around us.

The unscientific reader will naturally ask how it is possible to prove a negative proposition like the one we are here concerned with. Surely (he objects) all we can say is that no law governing the behaviour of the atom has been

---

[†] I wish to make it quite clear that I am not asserting the existence of a real indeterminacy in the physical world, or considering the philosophical implications of that view, especially in connection with the argument from efficient Causality. I am only urging that the acceptance of such a view does not dispense the mind from the duty of admitting the argument from Order.

discovered so far? To say that no such law *exists* would be to dogmatize; and it is the boast of science not to dogmatize. We are assured, however, that there is proof. Whoever has the skill and the patience to follow an intricate mathematical argument will find that the hypothesis of such a law existing leads to an absurdity. This, like all our modern arguments, ends up in an appeal to the expert. And since I am writing, not for experts, but for the profane, I have no option but to believe what I am told; or at least to argue on the assumption that what I am told is true. And, after all, even if the physicists were content to report that they could not trace the influence of any law in this one department of nature, it would be sufficiently strange. Wherever else we have questioned Nature closely, she has been more communicative. She does not always allow us to find out the reason for things, but she lets us trace a rhythm in them.

What do we mean, then, when we say that the law of averages steps in to neutralize the effect of indeterminacy? Let us take an example from common life for the sake of clearness. It would not surprise us to be told that every month almost exactly fifty umbrellas are left behind in the train at Crewe; the figures for the last six months have been (say), 49, 49, 51, 50, 48, 50. Now, in the case of the individual passenger there is no compulsion at work; there is no certainty, even, that he will forget his umbrella; a thousand factors are at work in his mind, and for all practical purposes you can say that the event is undetermined. Yet, when you examine the records of the Lost Luggage Office (I have not done so, all my figures are imaginary), the monthly ratio is for all practical purposes a constant. And if, like the modern physicist, you are grouping all the facts of nature under a set of mathematical formulas, it is not difficult to realize that the world as seen through the microscope might be a chaos of units that were subject to no law whatever, yet the world as seen through the naked eye might exhibit a merely statistical uniformity. If (to put the thing at its simplest) the visible world were nothing but a vast plain covered with sandhills, you might find that every hundred of these showed an average height of six feet per sand-hill, but the regularity of the observation would not greatly impress you. You would not say, "Here, surely, Infinite Wisdom has been at work!"

You would say, "The law of probabilities has been at work; odd, how it always does seem to work, when you take a wide enough area of observation! But we are accustomed to it."

But in fact the world of our experience has a uniformity which no law of averages will explain. The radium atom has, constantly, a different rate of decay from the thorium atom, and constancy of ratio cannot be merely statistical. If the authorities at Crewe told you that five out of six abandoned umbrellas was of the masculine type, you could not explain on merely statistical grounds why men are more forgetful than women. There is a *real* uniformity, it is evident, at work somewhere. Without taking into account the riddle of consciousness, and of man as a thinking being, we are still left dissatisfied with a merely mathematical diagram of it all. The Atom is in the news; its portrait, somewhat idealized no doubt by the skill of the artist, confronts us everywhere in the illustrated papers. Its likeness bullies the imagination, swims before our waking vision. But the world that was there before we ever invented the microscope is there still, and the Atom has not read its riddle for us. The Atom accounts for it, but does not explain it.

Rather, with each new discovery the mystery grows more profound; our wonder thrives on richer food than ever. The multitudinous variety of creation was baffling enough, viewed in itself; still more baffling, if we are to believe that it is educed from mathematical formulas, and statistical ones at that. It is as if we stood lost in contemplation of Chartres Cathedral, and a voluble guide plucked us by the sleeve to ask if we knew that it was all made up out of prefabricated material. It is no good trying to persuade us that the forms of things are merely a construction born of our own minds, which we read into our outward experience. The apple-tree is not a slice of reality arbitrarily cut out by ourselves, like the hundred sand-hills which we arbitrarily chose as the basis of our observation; it has organic unity of its own. It belongs to a pattern superimposed on the plain electron-stuff that all creation has for its background. We had admired the pattern as a mosaic; admire it more, now we find its medium to be a powdery gesso which, nevertheless, does not slip between the Craftsman's fingers.

Of such things, the layman writes with difficulty. It gives him a sense of irreverence, to be treating of such high matters with uncircumcised lips; he would fain be a scientist, so as to gain a juster appreciation, even according to our human measure, of the splendid craftsmanship in which God has revealed himself. He must take off the shoes from his feet; he stands on holy ground. Our age is in need of a great philosopher; one who can thread his way, step by step, though the intricate labyrinth of reasoning into which scientists have been led, eyes riveted to earth, by the desire to improve our human lot, the desire to destroy life, or mere common curiosity; one who can keep his mind, at the same time, open to the metaphysical implications of all he learns, and at last put the whole corpus of our knowledge together in one grand synthesis. He must be able to gaze through the telescope, to peer through the microscope, with a mind unaverted from that great Source of all being who is our Beginning and our last

End. He must be at once Thomist and Atomist; until that reconciliation is attempted, the pulpit and the laboratory will be for ever at cross-purposes.

There is a measure of truth in the old taunt of Lucretius and Voltaire, that man makes God in his own image. In turning your thoughts towards the Supreme, the Ultimate, you can choose, evidently, this or that aspect, this or that avenue of approach. And it has been a fashion with us, this long time, to concentrate on God's goodness as revealed in the moral order, not on his greatness as revealed in the natural order. To some extent, no doubt, we owe this to Immanuel Kant, and his metaphysical escapism. But perhaps even more to a certain softness in our way of looking at things, characteristic of a humanitarian age. The Old Testament has been under a cloud, more than we cared to admit, not only because learned men had been calling its veracity in question, but because its whole music was in an unfamiliar key; so much about fearing God, so little about loving him; references to his condescension that seemed only meant to underline his magnificence. The Evolutionists had been so emphatic about Nature red in tooth and claw that we fell into an unconscious sort of Gnosticism, treating the material world around us as the self-expression of some unnamed Demiurge we did not stop to think about; it was in our own moral instincts, our own upward strivings, that we found the best reflection of that Energy which permeated, somehow, the dull mass of created things. A Power, not ourselves, making for righteousness—that was the guarded formula under which Victorian moralism sought to find a constitutional status for the Ruler of the Universe.

I do not pretend that people of my own generation are likely to outlive the habit of mind which has grown up with us. We shall always tend, I think, to pray with our eyes shut, as if determined to reach God not by way of, but in spite of, his visible creation. But I wonder whether the Atomic Age will not take courage, from its new mental environment, to re-emphasize the lesson of God's greatness? *Rerum Deus tenax vigor*, the Church sings, when the sun's heat fades from afternoon into evening, as if to betoken that gradual dissipation of energy which is the time-fuse, they tell us, of the universe; we remind ourselves that it is the Divine power which holds all things together, holds them in being. And this new plaything which has come into world-history, atomic energy, is essentially the bond which holds the atom together, God's vicegerent in the work of conservation. Man liberates it at his own responsibility, not unaware that it is labelled "Dangerous." If it is prostituted to the service of destruction, that is our fault, who liberated it, not his, who energizes it. *Corruptio optimi pessima*; it makes a powerful solvent because it is such a powerful constringent; like human free will, it is a gift

God and the Atom

bestowed on us to make what we will of it; it sets before us life and death, blessing and cursing… We *liberate* it; be our machinery never so complicated and expensive, we recognize in atomic force something we have purloined, like the fire Prometheus brought from heaven. The same was true of the electric current, but not so obviously true; we talk about "generating" an electric current, as if we blasphemously claimed to be the parents of this noble bondservant that has been so useful to us. We shall never think of ourselves as "generating" atomic energy; the scale of the thing is so obviously beyond our human capacities. I like to think that these and similar parables will come to the rescue of the tormented human mind, and help it to dwell more effortlessly on the Power which holds us and all things in being, untiringly active in apparent quiescence, essentially beneficent, only capable of hostility in return for gifts misused… But perhaps the Atomic Age will be too mathematically-minded to have any use for parables.

Be that as it may, the Atomic Age will have, no less than ourselves, windows that open on eternity. The true lesson of the five proofs, as of all other proofs devised to establish the fact of God's existence, is that we see his face looking down at us from the end of every avenue of our thought; there is no escaping from it. All our metaphysics, play with word-counters and reshuffle our concepts as we will, must necessarily take us back to God. The doubts,

the hesitations, come only when human knowledge is suffering from growing pains, when we have not yet sorted
out our ideas and integrated, for the hundredth time, our
world-picture. Of that inevitability, our own heart-sick-
ness is the best proof. "Lo, all things fly thee, for thou fliest
me."

☙❧

VIII

An Alternative to Despair

I HAVE been suggesting the possibility that the Atomic Age will, on one particular side of things, find the approach to Natural Theology come easier. The picture of God as an omnipotent Creator will not, perhaps, seem remote or fabulous to a civilization which holds infinity in the palm of its hand. The world's debate, after all, is bound up with the contrast between mind and matter; the notion of a Creative Mind is not so difficult for the scientist to come by, when he feels that he, if he set his mind to it, could destroy matter—or at least introduce into the material of our planet an element of incalculable confusion. It does not follow that the Atomic Age will find it easy to believe in Providence; to maintain that attitude of optimism to which, amid all our doublings, we and our fathers have clung. There is nothing *immediately* repugnant to reason about the idea of the Universe as controlled by a hostile Power, a Juggernaut before whom we do right to bow, though we must not expect to get anything out of him. In practice, as I have suggested above, belief in God's existence does not usually survive the loss of belief in his Beneficence. But it does not seem impossible to imagine a world which would accept the doctrine of God's existence as a philosophical postulate without deriving from it either comfort or inspiration. A Deist world, but one in which Man would no longer be conceived as working a treadmill; rather, he would be entrusted with the appalling responsibilities of a Vice-Juggernaut. He would be a Prodigal Son, invested once for all with terrible powers to make or mar himself. Is this the prospect which actually faces us? The answer lies in ourselves; in our capacity (under grace, theology tells us) to keep alive the virtue of hope. Hope, more obviously than faith, derives its strength less from the certainties it feeds on than from an inner determination of the human will.

Of this, the Victorians and the post-Victorians were unconscious; theirs was an optimism which would not listen to the doubts of a Mallock or a Stevenson. That everything would come right, was in process of coming right, seemed to them axiomatic; the world had a roseate future, of which, had they stopped to examine the symptoms of their time, history offered a somewhat precarious guarantee. They multiplied, eager to people the millennium. It was only after 1918 that you heard the chilly doubt expressed,

God and the Atom

Is it worth while bearing children, to grow up in such a world as ours? Today, at this moment, those whispers have grown to an angry chorus. Optimism is at a discount; men clamour, not for happiness, but for security, and when a ship's passengers demand lifebelts rather than cabins, it is idle to pretend there is no danger of foundering. Heaven knows, there are more factors than one contributing to our present atmosphere of panic. But the imagination is most easily dominated by the threat of the Atom, the thing which has peeped out at us so suddenly and so enigmatically; flashed its message to us with a brightness above the brightness of the sun. Are we really expected to go on bothering about Utopia, when we have to allow for the possibility that modern weapons, used indiscriminately, might make of the world a Utopia in the literal sense, a Nowhere?

To all these hesitations, there is a Christian answer close to our hands; a crisp, ready-made answer which silences all discussion. I have indicated the lines of it in an earlier chapter, but it is time we set it forth in more detail. It is this. The Christian virtue of hope has nothing whatever to do with the world's future. As it was preached by the first apostles, it meant nothing more or less than a confidence on the part of the Christian that he or she would attain happiness in a future life. The world about them was perishable, and doomed to perish—perhaps in a very short time. The agonies of its dissolution might have terrors for the wicked and the worldling, none for the believer. The prospect was a warning to us, of course, to live in a way worthy of our vocation, but apart from that, whether the world lasted ten days more or a thousand years more was no concern of ours. Nor was that way of looking at things confined to the first ages. St. Gregory expects the world to come to an end quite soon, and appeals to an *immutatio aëris*, a change in the atmosphere, discernible (but not, I think, elsewhere recorded) in his day. Again and again you find religious revivals accompanied by a belief in the impending dissolution of the world order. What does it matter, then, to us Christians, if we find the children of the world, like the fools they are, playing with fire? Let them destroy or devastate the planet; we, the elect, we, the remnant, see only a liberation from the weariness of this mundane existence, where they see calamity. Our hope lies, not in this world, but in the next; if it is true that the portents of our time give promise of a general conflagration, so much the better. It is doubtful if a world that has forgotten God, as ours has, can deserve or even desire a better fate. We always told them that their dream of a wiser and happier age was doomed to disappointment; now, perhaps, they will see that we were right.

That (I say) is a possible attitude; and there is little doubt that it will be a common attitude in those restricted circles where the Apocalypse is eagerly scanned, and the Pyramids measured, in search of assurance that Dooms-day is not far off. It must be confessed that there is something new in the idea of the general conflagration being caused by man's own action, but no ingenuity will be spared in searching for Biblical predictions in this sense. We are not concerned, however, with these professional students of prophecy. What is more significant is that there is a tendency among people of more conventional piety to take refuge in the other-worldly attitude. The partisans of Utopia have so long been dinning it into our ears that science is the beacon-light which points the way to universal happiness, that we should be scarcely human if we were not tempted, for once, to get our own back. But it may be doubted whether this type of Christian *schadenfreude* is the best we can do. When you have lost your way and are asking for directions, few things are more annoying than to be given, by some local Good Samaritan, an exact account of where you went wrong. Our contemporaries, and posterity if there is any, will be more grateful to a Christianity which can offer them some message of encouragement.

With that in view, I must ask leave to consider, with more attention than is usually given to it, the doctrine of hope; a doctrine baffling and full of paradox. Hope is something that is demanded of us; it is not, then, a mere reasoned calculation of our chances. Nor is it merely the bubbling up of a sanguine temperament; if it is demanded of us, it lies not in the temperament but in the will. Indeed, we can hardly doubt that the Christian is at his best when he is, as we say, "hoping against hope." Hoping for what? For deliverance from persecution, for immunity from plague, pestilence, and famine, from worldly discomforts in general? No, for the grace of persevering in his Christian profession, and for the consequent achievement of a happy immortality. Strictly speaking, then, the highest exercise of hope is to hope for perseverance and for heaven when it looks, when it feels, as if you were going to lose both one and the other.

One thing stands out luminously from the study of all the mystical authors—that the higher stages of the spiritual ascent are prefaced by a period, sometimes by a long period, of desolation; the soul feels as if it had been abandoned by God. And there have been saints—St. Francis of Sales is a well-known example—who in that time of purgation felt certain, not feared simply but felt certain, that they were doomed to eternal loss. What, then, should be the posture of a soul which is submitted to this terrible ordeal? The answer given to that problem by the Quietists was perhaps the chief reason for their condemnation, at the end of the seventeenth century. They argued that resignation to God's will was the most indispensable of all duties; that perfect love of God meant loving him for himself, without expecting to be rewarded for it; therefore if you felt certain that God had doomed you to perdition, your duty was to acquiesce in the sentence. After a religious controversy perhaps unsurpassed in history for the brilliance with which it was conducted—Fénelon led on one side, Bossuet on the other—the Church decided that you had no right to make such a gesture. You might, indeed, by a kind of pious rhetoric, profess to God that you would be prepared to suffer eternal loss if, *per impossibile*, that was his absolute will for you. But, since it is certain that he does not refuse grace to the man who does what in him lies, you must not give credit to the infernal whisper which told you that you were beyond redemption. In the midst of your despairs, it was your duty to go on hoping.

But can a soul really *hope*, it may be asked, when the whole mind is overshadowed by the conviction that there is no hope at all? The best answer to that question is implied by a well-known passage in the *Imitation of Christ*. "There was a man that was tossed ever to and fro between hope and fear. One day, overcome by melancholy, he went into Church and threw himself down in prayer before one of the altars. Ah, he thought to himself, if only I could be certain that I should go on persevering! Whereupon he heard the Divine answer in his heart, And if thou wert certain, what wouldst thou be doing? Do now what thou wouldst be doing then, and thy anxieties will vanish." *To go on behaving as if we hoped* may be, for some of us, at bad times, the nearest approach we can make to hoping. But if we do not make at least that effort, there is grave danger that we shall really lose our souls, by taking leave to treat them as if they were lost.

Let me repeat, to safeguard my own orthodoxy, that hope in the theological sense is concerned only with the salvation of the individual believer, and the means which will help him to attain it. No Christian must accuse himself of the sin of despair merely because he finds himself thinking that his country will be beaten by its enemies; nor, for that matter, because he finds himself thinking that the world is very evil, and probably on the verge of blowing itself up. But there is a kinship between the theological virtue of hope and the endearing qualities of Mark Tapley; the law holds good, even in human affairs, that you must go on behaving as if you hoped, if you would save your *moral* from crumbling. You are beaten when you throw up the sponge.

If a world-disaster is to fall upon us almost immediately, there is no more to be said. But this is, after all, only a possibility; it depends on the powers of the atom, which the scientists do not pretend to know; it depends on the future behaviour of human agents, which no one can possibly know. In the alternative, is it fantastic to suggest that the world, under God's Providence, has come in for one of its bad times, just as the mystics tell us that they, at such and such a stage in their careers, came in for a bad time? At first, these people tell us, everything went smoothly with them, God's face smiled on them, and they were in danger of presumption—of thinking that they could sit back and let God do the rest. Then, by a swift change in his dealings with them, they fell from crest to trough; God seemed to have deserted them altogether, and nothing had savour for them any more, in this world or the next; they could hardly doubt that they had been cast away for ever. They were in danger of despair; of thinking that nothing they did made any difference. Perhaps this later experience of theirs was a punishment for their earlier presumption; they did not stop to enquire. One thing was certain, the ravages of presumption would not be mended by despair. They must go on hoping; registering hope (so to say) with every fibre of their wills.

If I indulge the fantasy that the world, in its present uneasy state, may perhaps be going through the same sort of experience which holy people have often met in the course of their purgation, I must not be understood to suggest that there is any real theological parity between the two things. All that I say here must be taken as a kind of Platonic myth, eking out with imagination the gaps in our knowledge. We have no knowledge, after all, of what God means the human race to be, or of what he means to happen to it, in the large. We know that, for each of us, the particular historical and social setting into which he or she was born forms part of the raw material out of which an eternal destiny is to be shaped; the business of religion is with the individual. But surely it is difficult not to believe that the history of the human race *as a whole* is something which God wills, and wills with a view to some end. The race, like the individual, has its opportunities, its ups and downs, its vicissitudes. But it is impossible to say for certain what God's plan is for it; whether (for example) he means it to grow happier, as the wisdom of the ages accumulates in its memory, or more wretched. Scripture, tradition, reason afford us no ground for prophesying one event rather than the other. The study of history shows you a fairly steady advance in material comforts, in scientific knowledge, and so on; but to determine whether man is happier in one century than in another is a fool's task. We can only assume that "through the ages one unending purpose runs," and leave that purpose obscure. Because it is obscure to us, we eke out the gaps, as I say, by mythmaking in the Platonic manner. Here, then, is the myth I propose.

God means the human race, or some large part of it, to grow towards himself, to become more perfect. Not necessarily in the sense that, as this movement goes forward, a larger number of souls will attain heaven in one century than in the century which went before it. He makes, we must believe, infinite allowance for our opportunities; and, for all we know, there may be thousands of souls dying at peace with him in some environment which seems utterly hostile to the reception of his graces. No, but, viewed in the bulk, humanity will become more what he would have it become. Nations will be more at peace with one another, governments more just, public opinion more humane and more enlightened—all the rosy dream of the Victorians will tend to come true, and at the same time, a thing they wasted little thought over, religion will progressively come into its own. It will be a reign of Christ the King; not a millennium, but a progress towards the millennium. All this may be—I do not say is, but may be—God's effective will for the human race, destined to reveal itself more fully in the centuries that lie before us.

Assuming that this myth has any value, should we not expect to see the same alternations of light and darkness, of consolation and desolation, in the experiences of a race heading towards perfection, as in the experiences of an individual heading towards perfection? And should we not expect to see, in the race as in the individual, a tendency towards presumption when the good times come, a tendency towards despair in the bad times? And should we not rightly conclude that, for the race as for the individual, the secret of progress lay in fighting down those tendencies, each as it came?

RONALD KNOX

Nobody looking back on the years between Waterloo and the outbreak of the first World War can doubt, I think, that they were good times. I am not speaking of material prosperity, though that had much to do with it. The essential characteristic of the period was its optimism; the unconscious assumption that everything was going right, and going in the right direction. You could almost hear the dawn breaking…. We ask ourselves now, Was all that an illusion? Or was the world of that day really well on the road to happiness; and has malignant fate, or blind chance, wrecked the machine? We are tempted to fall back on old pagan notions, of a Divine envy which cannot brook the sight of man enjoying uninterrupted prosperity, of a fatal human folly which goes too far in its determination to exploit success, of a Nemesis which automatically recoils upon the rashness of the attempt. Would not Aeschylus, would not Herodotus have pictured Heaven as grudging mankind the aeroplane and the wireless, mankind as growing weary of these toys and asking for something fresh, atomic force as the fatal answer to that prayer, a Nemesis sent to rebuke us for our impiety?

It is for that pagan myth that I would beg to substitute a more Christian myth, of the human race growing towards God, but by means of trial and error. One fact emerges plainly from any study of the mystical writers—that there is a kind of false start which often makes beginners think they have travelled far along the ways of the spirit, when in fact they are only beginners. It is like those treacherous warm days in early spring, which make us give up fires and discard our winter wear. The account generally given of it is that God is inviting the soul to turn its back on the world by giving it some foretaste of what union with himself means; but only a foretaste. There could be no greater mistake than for the beginner to imagine he can neglect precautions, rest on his premature laurels. Many souls err at this point through presumption, but not necessarily with fatal effects. They imagine they have already reached a state of union, that their present happiness will remain for ever undisturbed, and they are all the less prepared for that time of difficult purgation which is waiting for them round the corner. Is it possible that the world of yesterday was something in the same position? That it mistook some fleeting foretaste of a civilized world-order for a civilized world-order already in being, already firmly established? We did presume, no question of it, upon our long experience of prosperity, but there is no reason to think the presumption was fatal. The mischief of it was that it left us ill prepared in mind for the searching purgation which was to follow.

During the last thirty years, most of us have felt that the spirit of optimism, to which we had so long been accustomed, was running down. It was as if a switch had been turned off at the main, and a deceptive glow still lingered, to mock us with a transitory illusion of warmth. We saw the lights of freedom go out one by one, and tried to persuade ourselves that it was only a phase, that men would grow tired of false philosophies, that the balance of power would hold; but we knew it meant war. We faced war, and went through with it, but depressed by a consciousness, unexampled in our annals, that victory would bring with it no solution. We were amputating a gangrened limb, too late. And at the moment of victory, a sign appeared in heaven; not the comforting Labarum of the Milvian Bridge, but the bright, evil cloud that hung over Hiroshima. In this sign we were to conquer.

It hung like a mocking question-mark over the future of our race; It hung over a world darkened by the shadow of famine, in which a great part of mankind seemed threatened with hopeless exile, in which civil feuds tore at the heart of every country, in which brute force, under a pretext of cowardice, was scheming everywhere for its own aggrandisement. Tacitus would have us believe that when the Roman armies stormed Jerusalem, "a cry went up, that the gods were leaving the city, and a great stirring was heard as they left it"; was God leaving the world to its own devices, wearied at last by our long record of infidelity?

The myth tells us that this gloomy presentiment of ours is nothing to be surprised at, and at the same time that it is not to be trusted. It is precisely the experience of souls which, venturing outside the common paths of prayer, have suffered Divine things, that God does seem to hide his face, does seem to cast us off altogether, at those times when consolations are to be left behind, and we must enter upon a stage of purgation. Yesterday and the day before he was making trial of us, to see if we could stand prosperity without giving way to presumption; now he is making trial of us, to see if we can stand adversity without giving way to despair. The attitude he wants to find in us is Job's attitude, "If we have accepted blessings from God's hands, should we not bear it with courage when he sends us ill fortune?" It is possible that the crescendo of disillusionment which has marked these last years is only, after all, a purgation. He wants to elicit the virtue of hope in us, by making everything seem hopeless.

However discouraging, then, we find the portents of our time, we are not necessarily justified in abandoning the old instinct we had, that there is a Providence watching over human affairs. It is not necessary to conclude that the world is hurrying to its dissolution, leaving us (in a theological sense) to raise the cry of "Sauve qui peut." It may be that mankind is being called upon to exercise the virtue of hope; and, if so, Christian people must think twice before they abandon themselves to the luxury of world-despair; before they wash their hands of our communal guilt, and betake themselves, singing "O Paradise, O Paradise!" to the hilltops. We shall do better, I think, to help man the pumps of the labouring ship, and let the world see that hoping is one of our specialities.

St. Paul has some useful advice (1 Thessalonians 4:11 and 12, 2 Thessalonians 3:11) for the sort of people who give up taking any interest at all in worldly matters because they think the world is just coming to an end. And he himself, shipwrecked on his journey to Rome, preserves an attitude which is worth studying. He has little to hope for, and the journey from Crete was undertaken against his own advice. Yet it is he who restores the *moral* of the dispirited crew. Salutary enough, when the world thought all was going well with it, for an occasional protest to be registered, "Why exercise yourselves so much over the future of this transitory world? Some day, tomorrow perhaps, it will all be reduced to ashes." It is otherwise when the first whispers of panic are spreading abroad; when the old are lamenting the disappearance of all the landmarks they knew, and the young are asking whether there are homes and jobs for them in the Europe they have fought to save. Then it is the part of prudence, and perhaps of common humanity, to keep cool heads and offer what encouragement we can. It was galling, no doubt, to be told (as we were told till lately) that our Christian world-picture was unscientific; that the earth would last for millions of years yet, and only become uninhabitable through the slow cooling of the sun's heat. Now that the general conflagration is, by some accounts, almost in the sphere of practical politics, in heaven's name let us have the grace not to say, "I told you so."

If you believe in Providence, you believe not merely that the course of external nature is directed towards a purpose, ultimately beneficent; you have to believe the same about human actions. Quite apart from his inspirations in the order of grace, God overrules our human designs; manipulates the tyrant as much as the tornado. The world of yesterday consented to abandon its belief in Providence, but not consistently. Full of Darwin and his survival-values, it refused to believe that external nature was ordained to any purpose whatever. But it continued to hold that human history was a history of steady progress, ascribing the fact not to any Divine interference, but to some innate striving for goodness in the human species. Now, for the moment at least, that doctrine is at a discount. Nobody believes that the world, at this moment, is getting steadily better, except the people who are content to single out whatever tendencies are in fact carrying the day, and to label them "Good." Most people who are capable of thinking, and are not deceived by wish-thinking, agree that the world is in ferment, and the Atom alone knows what is going to be the result of it all. What are we Christians to tell them?

RONALD KNOX

Are we to tell them that they were wrong, and we always knew they were wrong? That all the progress we boasted of yesterday was an illusion and an irrelevance, that the kindliness they preached was only the sentimentalism of the well-to-do, that the world's peace, when there was peace, was dictated by financiers for their own ends, that the sanctity we once attached to human life was inspired only by a common fear? Shall we point to the present condition of the world to justify our reading of the facts; assure them that the world is a vale of tears, and they were fools (to put it bluntly) if they ever thought anything else? Shall we maintain that politics, always and everywhere, is a dirty trade, that publicity is all lies, that any hope of betterment on this side of the grave is an idiot's dream? Shall we treat them as we might—not very wisely—treat a drunkard who was down and out, by reading him a lesson on the evils of drink?

What I have been trying to suggest is that there is a more hopeful and a more human attitude open to us, one which gives more credit to the imperfect strivings of an earlier age, one which assumes less detailed knowledge of the Divine purposes. We can say, in effect, "Do not be surprised at the fiery trial which has come upon you. God's way, to judge by his dealings in ordinary human lives, is first to encourage us by success, then to discipline us by failure, now fortifying us against despair, now warning us against presumption. It would not be unlike his methods, so far as we know anything of his methods, to let civilization advance up to a certain point in the right direction, without confirming it in goodness, without guaranteeing it against the possibility of relapse; then, to withdraw his assistance from us for a time, and let us see how helpless we are without it. We cannot be sure, but that is, probably, what he is doing at this moment. He is preparing us for still greater advance in the future by letting us go through a phase of purgation which may, for all we know, last out our time. But he still rules, not Nature merely, but man's will, and if he has given us a new weapon which is capable of burying us all under calamity, he is perhaps only testing our wills, to see if we have the decency to forswear the use of it. Even when he put Man in Paradise, he planted a tree there from which it was Man's business not to eat."

No need to add that there is another side to it all. Resignation, abandonment to God's will, has to be preached as something more obviously necessary than ever. But we must not, by a kind of cosmic Quietism, exalt resignation to such a point that hope has no longer any meaning for us. We must steel ourselves for either event, in the spirit of the old prayer: "Lord, if thou wilt have me be in light, blessed be thy name; if thou wilt have me be in darkness, blessed be thy name. Light and darkness, bless ye the Lord."

IX

An Alternative to Decadence

WE ARE dealing, let us tell ourselves again, with mind-pictures. Three especially concern us. The first, barely possible to the imagination, is that of the Atom waiting to explode at a moment of its own choosing, without any reference to law. The second is that of a city enveloped in darkness, full of men fleeing, and burning as they flee. The third is that of a control released, now at the heart of the atom itself, now in mid-air over the doomed city. Faith feels a twinge at the first, hope at the second, charity at the third. I have tried to show that mystery, in nature as in supernature, ought to strengthen faith; mystery is its proper food. I have tried to show that hope, taken in a wide rather than in a strictly theological sense, is a quality we Christians should be prepared to advertise for export. What of charity? Can we expect the rising generation to submit tamely to controls, whether their sanction be human or Divine, when the picture modern science feeds their unconscious minds with is a picture not of control, but of release?

I confess that this last appears to me the most disturbing element in the situation. I almost wish the Atomic Bomb had been dropped on Germany. For the Germans, be they what they may, live on the dregs of a Christian civilization; Rosenberg's Walhalla-nonsense was too obviously synthetic to catch on. If Berlin had been wiped out, the Germans would only have said, "See what poor Christians these English are!" But the Japanese are more likely to say, "See what hypocrites these Christians are." In the Pacific, you have a direct conflict between two cultures. The Japanese admire our Western efficiency—they know a good deal, for instance, about radio-activity. They do not particularly admire our Western morals, or understand why we preach self-sacrifice, and yet allow ourselves to be taken prisoner instead of committing suicide. East and West will be at cross-purposes more than ever; the Atomic Bomb will be regarded, out there, as the symbol of British-American civilization. Is it to be the symbol of British-American civilization? Do we mean, on principle, to relax all controls?

We must remember, though, that it was the old explosives, rather than the new, which made us think of detonation as an instance of hidden possibilities suddenly realized, of potentiality passing into energy. We even christened dynamite after the Greek word for potentiality. The train without the match, the match without the train, was inoperative, remained dully quiescent. We took that for a symbol of our native inertia, the apathy or cowardice which got nothing done until some firebrand came to set it ablaze, "the unlit lamp and the ungirt loin." We diagnosed, some of us with impatience, some of us with apprehension, the presence of explosive elements in society which only needed the arts of a skilled agitator to put the match to them. Youth frustrated of its aims found a symbol ready to hand in these unrealized capacities for revolt. To the revolutionary, dynamite became a kind of sacrament, symbolizing what it effected.

But the natural symbolism of atomic power is something quite other. In using it, we think of ourselves, not as reducing potency to act, quickening *a* dead thing into life, but merely as liberating a force which is already there, *and already at work*. (Science bids us think of energy in the atom, too, as only potential. But we are not concerned, here, with an accurate scientific description of the facts; it is the popular representation of them, the popular language about them, that gives them their symbolic value.) We are accustomed to think of energy as implying motion, but there is a sense in which it can be active without producing any effect on our senses at all. "The arch," say the architects, "never sleeps"; its agelong permanence is due, not to mere inertia, but to a continual balance of strains. The weight it bears is continually trying to thrust the two arms of it outwards; only the counter-strain of the buttresses keeps that tendency at bay. Knock one of the buttresses down, so that the arch falls, are you reducing potency to act, or are you liberating an energy which, for centuries has been silently at work? So with the atom; the energy which dwells in it is in a sense employed already; all we have to do is to bring it into play, to liberate it.

The word "liberation" has undergone, in our time, a melancholy change of meaning. A country can be liberated, irrespective of whether its inhabitants wish the thing to happen or not. And the vast research, outlay, and apparatus which were involved in the liberation of atomic energy would suggest that the wishes of the Atom (if I may so personify it for the moment) were not consulted. I am not trying to suggest that splitting the atom is something "contrary to nature," and therefore wrong in itself, although I can imagine a scholastic discussion to that effect being staged with much ingenuity. I only mean that it would be a very far-fetched piece of metaphor if we represented the atom as a damsel in distress, immured in a dungeon for centuries and emancipated at last by the paladins of science. Nature, by all accounts, could not have made a more becoming show of maidenly reluctance in yielding up this, the last of her secrets. In fact, it was necessary to bombard the atom before we could secure its consent. And, so far from adopting it as the totem of the explosive, revolutionary forces in the world, we should be better advised to blazon it on the arms of what are called "the peace-loving nations." If it is destined to involve us all in disaster, at least it did not wage a war of aggression. It had to be attacked, and violently, before it consented to hit back.

But indeed, this is no subject for frivolity. What I want to make clear is that if, in the new age, we mean to borrow a pattern for the architecture of our lives from Nature, we must think of atomic energy in its normal setting; not as carried away by Man at the saddle-bow, but in its virgin state, fulfilling its primary function. You can use an elastic band to flick pencils about, but its primary function is to hold things together; and such, it seems, is the primary function of atomic energy. Why should not the material, here, be the image of the spiritual world? The atom is the unit of one, the human soul of the other. If, in the atom, there is a unifying principle which integrates those infinitesimal constituents into matter, the human soul, too, needs a unifying principle to integrate it; to save it from becoming a mere welter of impulses and urges. If, in the atom, this unifying principle is not mere *vis inertiae*, but a force which man borrows to blow towns to pieces, then we must not expect the integrating principle in a reasonably lived human life to be mere *laissez-faire*; it is the strong man armed that will keep his goods in peace. If, for count-less centuries, man was wholly ignorant of this secret force that lurked beneath the outward appearance of things, it is not surprising that men should neglect, and even ignore, this integrating principle in which, nevertheless, lies the whole business of living.

Something of the same mystery which surrounds the atom surrounds human personality. The atom is the smallest thing in nature of which you can say, "This is so-and-so," giving the name of the substance to which it belongs; what lies beneath can be known only by quan-tity, not by quality; qualitatively, it is indivisible. Human personality is indivisible in a more strict sense; its facul-ties of memory, intellect and will are not component parts, which added together make up the whole, not yet can you satisfactorily think of them as mere belongings, tacked on to the personality from without. A person cannot, strictly speaking, become more or less of a person; when we talk of a "split personality" we mean only an unexplained gap in the memory or in the will; that person, and no one else, will always be the subject of that person's experiences. Yet there is a mystical sense in which the man who neglects his own spiritual life becomes *less of a person*. He identifies himself first with this motive, then with that, the two being quite inconsistent; throws himself into pleasures of which he is ashamed or repentant before the day is out; is carried away by violent prejudices, which vary from one month to the next; worries about unimportant things that lie in the future; complains about minor discomforts. He is all over the place; there is no unity in his moral being. He needs, we say, to integrate himself; where is he to find it, this atomic energy that will slip an elastic band round the pile of untidy fantasies that makes up his life?

It is no use to say that he needs self-control; seldom did any man set about controlling himself for the sake of controlling himself, without starting or at least finishing up as a prig. He must be inspired not by the name of a process but by a real spiritual force; that is, a motive. Plato called it "justice"; it was justice, he said, in the commonwealth that made all the three classes of the community respect one another's rights; it was justice in the individual that held the balance between the three corresponding elements in Man's soul. But the word is used in a sense foreign to us; it is difficult to give it any more concrete meaning than civic sense in the one case, moral balance in the other. If we can find no name for it, this restraining principle, the reason seems to be that no such principle exists *in us*. Men, like atoms, behave in a predictable manner when you take them in the mass, but individually, if left to themselves, consistency is not to be found in them; they are the sport of caprice. (I do not say anything about Divine grace, because that is not to our present purpose.

RONALD KNOX

The office of Divine grace is to perfect, not to replace our human processes, and it is these that concern us.)

The plain truth is, that the motive has to come from outside. To be eaten up with love or admiration for a person, with loyalty to a cause, with conviction, even, about a system of thought, often has an integrating effect on a man's life. Unfortunately, if these belong merely to the natural order, they either produce consistency of conduct only in a particular field, or produce it only for a limited time, till the romance has faded. What religion does, in its pure form, is to provide all these three motives at once. I say, in its pure form; for there is a regrettable and fairly general tendency among religious people to live by a code. Man is so much a creature of his environment that he is always ready to fall back on living by a code; and the code of religious people is usually somewhat, but not startlingly higher than that of their neighbours. Hence arises much scandal, both to the Pharisees and to the little ones; religion looks, from the outside, not much better than a sham; feels, from the inside, a precious tame business. That is not surprising, where a code has supplied a merely conventional substitute for that terrific astringent force religion was meant to be.

In itself, it is a terrific *positive* force; it inspired the Saints to prodigies of love, bade them go out and kiss the sores of lepers. If you do not accept those stories, consider

how, like the atomic force, it becomes dreadful when it is
deflected. Fanaticism is not madness, it is religious zeal
miscanalized. The cruelties of Covenanter or Camisard
became possible not in spite of, but because of, their religion. Yet, amid the whirling stream of impulses that is our
nature, religion becomes true to its name; it binds. That
is true of the other motives we mentioned as integrating
factors in human life. A man will give up his darling vices
because he is in love, will ruin his prospects for the sake of
a cause, will go without food and sleep because he is too
busy arguing to attend to them. You cannot have everything; the integrated person is clearer about his notion of
the priorities than we others, that is all. It is for the ideal
statesman to know exactly which claim must have the priority here and here, when he knows that it is impossible
for his fellow citizens to have a large army, and houses,
and goods for export, all at once. It is for the ideal man to
know what sacrifices he has to make, and make gladly, if he
would achieve the end he has in view.

Not that religion, when it becomes a strong force in a
man's life, really involves these nice, these laborious calculations about means and ends. Rather, he has a pattern
before his eyes, and can see at a glance what colours match
it, a melody in his ears which would make this form of
action or that—he cannot quite tell you why—a discord.
Why did St. John Vianney desert from Napoleon's armies?

If he did it to spare himself, it was the only time he was known to spare himself in seventy-three years. It did not fit into the pattern…. At a quite different level, readers of Maurice Baring's *Passing By* will remember how the heroine becomes a nun and refuses to marry the man she is in love with, when her husband has died, but died partly through her fault. "If she married Y., that would make a legitimate harmony certainly. But her very feeling for the *full* harmony of life would make it impossible… for her to use X's death as a means for doing rightly what she had meant to do wrongly…. Within the harmony of her marriage the memory of that discord would always be present." Never more than when you are in contact with the supernatural, *le mieux est l'ennemi du bien*.

As I say, most of us would have been well content if, with the Atom Bomb poised over Hiroshima, the Allied Nations had said, "No, we are not going to drop it. You Japanese fellers probably won't understand why, any more than you understand why English people play cricket, or Americans play baseball. All you can say is that it's part of our code, part of our *bushido*; it's one of the things which aren't done." But some of us would have been still better content if the Allied Powers had said, "No, we are not going to drop it. We are not going to unleash against you the strongest material force yet discovered, because there is a still stronger spiritual force which binds our hands.

God and the Atom

Right or wrong, that action would not have fitted in with the pattern of life we Christians try, heaven knows how unsuccessfully, to live by; it would be false to the rhythm which has made European history." Somewhere, I think, for us as for St. John Gualbert, a Figure would have bent down from a Cross in salutation.

I know it sounds absurd to suggest that a Christian gesture should be made on an international scale. Human beings here and there may rise superior to their environment, but not masses of men; humanity in the bulk has to be content with a code, and, in war-time, not a very high one. Let us take comfort at least from the scientists' assurance that the uniformity of nature itself is only statistical; at a lower level, things become unpredictable. Just so we, human atoms, are not bound to take our cue from humanity in the mass. There is nothing about the behaviour of modern nations (that is, in many cases, of the gangs that control them) which is calculated to edify the men and women of the rising generation. The leading articles of the world's press, if you sit down to examine them dispassionately, and read a little between the lines of them, seem dictated by only two considerations, Cowardice and Grab. Nor does mass observation of our fellow creatures— at a railway station, for example—give us a much more cheering picture of the human average. The bond of the war-effort once relaxed, we seem to have disintegrated; it is a matter of general testimony that our tempers have got worse; we scramble, with less shame than formerly, for all we can get. We hurry over news of starvation on the continent of Europe, in our eagerness to make sure that *our* rations are not to be docked. We submit to innumerable forms of control, but no longer with a good grace; no longer with the consciousness that we are all one hive; we are the unenthusiastic victims of Whitehall. Disintegration has set in; and dare we hope that the lives which are being formed in such a school will be integrated lives?

If I express a hope of the kind, I shall no doubt be accused of wish-thinking. If Plato doubted whether the ideal man could be produced (except as an occasional freak) until you got the ideal state, who am I that I should bid my fellow citizens rise above the level of their age? I know. But is it necessarily paradoxical to suggest that a young man of generous instincts might tend to react from his surroundings; might become conscious of the need for integration in his own life by watching, not with cynical detachment but with sympathetic human interest, the disintegration which he sees around him? After all, to be brought up in the kind of society where there is always rather more than enough to go round; where, in consequence, the ordinary citizen is effortlessly kind, effortlessly generous; where peace abroad, relative content at home, make for a prospect of security, and there are prizes even for the unambitious—all this does not necessarily assist in the formation of a mind startlingly Christian. Youth does not suddenly slap its thigh, and say, "Heavens, how good all these people are! It is up to me to go one better!" On the contrary, you are more likely to get a crowd of amiable non-entities, queuing up to find entrance by the broad gate; it is the thing to do. The young Englishman in particular, trained in the habits of the herd, is very apt to say that he "can't see much wrong with" the society which surrounds him, unless he gets a considerable shock to jolt him out of his composure. Is it not possible that the sight of a disordered world, in which his own country is making a difficult bid to keep up its position of importance, is the very thing needed to throw him back on himself and make him ask, "Have I *got* to be like everybody else?"

There is this, too, to be considered; that if you make a man less of a citizen he tends to become more of an individual. Nature takes its revenge; if he cannot find elbow-room in his civic environment, he is all the more eager to make the most of his own personality; a slave by accident of external circumstances, he will be ruler in the narrow kingdom of his soul. That is why, notoriously, the Stoic philosophy flourished in the early Roman Empire, when men who had not forgotten the taste of liberty found themselves condemned to be the unwilling subjects of a world-state. The whole preaching of the Stoics was self-integration; they had an elaborate ascetic code, which gave to every department of human endeavour its suitable priority; they exercised a rigid discipline over the minds and senses. So, still earlier in history, you get the classic protest of Diogenes against Alexander; the Macedonian armies have extinguished political liberty in Greece, but they cannot extinguish the liberty of the soul; the Cynic in his tub is a monarch still.

Mankind has abdicated, within a life-time, all those rights which we used to include under the notion of political liberty. To be free, in these days, means at best to be bullied by your own fellow countrymen rather than by foreigners. Commonly enough, you do not even enjoy the privilege of voting at an election unless you are an active supporter of the only party which is allowed to run candidates. Your property is yours only on sufferance; you must hand on no traditions to your children except what the State approves. Even where we still hold out against ochlocracy, and enjoy democratic institutions, the shackles of State control tighten round us daily; inspected here, directed there, we find the area of individual choice continually shrinking, our lives increasingly conditioned from above. To discuss the rights and wrongs of all this is no part of my purpose here; all I am concerned to point out is that we are not the *men* we were; in all those daily contacts which make citizens of us, our personalities are being ruthlessly abridged. And where that happens, it is the instinct of a generous nature to build up personality from within, because it has no opportunity to develop outwards.

If religion, as Lenin said, is the opium of the people, he himself has done his best to make drug-fiends of us all. We *must* find some outlet; blame the escapist attitude if you will when it means running away from facts, but not when it means recognizing the facts, and making the best of them. One thing I take to be quite certain about the Atomic Age—the police are going to have the best the street fighting. Where the means of destruction are widely distributed and easy to come by, there is some chance of popular indignation dethroning the tyrant. But it is likely that our new fighting weapons, depending on so much apparatus of laboratories and factories, will be jealously kept beyond reach of the public. Nothing but a palace revolution will be possible, and the ordinary man, deprived of any opportunity to take an effective interest in the future of his country, will be driven in on himself; sometimes, let us dare to hope, with salutary results. It was said of a man suffering from an unexpected illness, "God stretched him on his back, to give him time to think." Let us dismiss the ungracious image of the world as a slave in a dungeon; let us call it, instead, a sick world that has gone to hospital. The doctors, some of them, have rather kill-or-cure methods; the nurses have not always a fortunate bedside manner; but anyhow, here we are, with time to think.

What meditations, then, am I suggesting to the young man anxious to make the best of tomorrow? Nothing very profound. Let him take Hiroshima for his starting-point, and reflect that if the decision, whether to bomb Hiroshima or not, had lain with a single man, not bound to consult anybody's interests except his own, Hiroshima would probably have been spared. The decision that was taken does not reflect the individual will of President Truman or of any other single human being; it was the work of statesmen who, quite rightly, acted as the representatives of the countries which had given them high office. They had to consider the misery, the loss of life, involved in prolonging a modern war twenty-four hours beyond its necessary limit; they had to consider the feelings of innumerable citizens who, sooner or later, would have to hear about their deliberations; they had to consider the reactions of Allies who were not present.

It was not men, in fact, but nations that condemned Hiroshima to suffer. And a nation at war has difficulty in keeping its hands clean, because it cannot keep its hands free. A whole bundle of human interests is concerned, whenever a decision is made, and these cannot be neglected. Generosity, the gesture of claiming something less than your rights, is to be found in the individual, not in the group.

I have said "a nation at war"; but indeed, the state of emergency which war creates is slow to die down, perhaps never dies down altogether in diplomatic affairs. We are entering on a long period of bargaining and chaffering between the great world Powers, in the course of which many fine phrases will be used; but when the Peace Treaty comes, it is unlikely that we shall find any single award that was either made or accepted on *demonstrably* unselfish grounds. If the frontiers of Ruritania are enlarged, we shall assume at once that this was not done from any desire to make the Ruritanians happy, but because it suited the book of the Power which won the day at the conference-table. Sitting round that conference-table, the delegates will be thinking of many things—the aspirations of people at home, the solid interests of their respective countries, the elimination of future dangers; there will not be much room for generosity in their thoughts. Once more, they are not men, they are representatives of mobs of men. And the mob is not generous. If any doubts are felt about the truth of this item in the meditation, a visit to the nearest railway terminus will set them at rest.

But man, the atom of humanity, can choose to be generous if he will; he has choice. There are warring elements in the make-up of his own mind, impulses craving to be satisfied, prejudices ready to warp the judgement, self-love in a thousand forms, even capable of masquerading as high-mindedness, and all the rest of it. But these clamorous voices have no individual being; the man himself is the unit. He may, as the proverb says, cut off his nose to spite his face if he wants to; it is *his* nose, *his* face; no alien interests are concerned. Most of us spend a lot of our time doing something not unlike what the proverb indicates; ruining our health for the sake of our pleasures, our peace of mind for the love of chewing a grievance, our reputation for the sake of gratifying a grudge. But we know really that we want to claim the atom's privilege of integrating our lives.

Only a religious force will effect that integration. Every religion demands something of sacrifice, at least of those who mean to live up to it; there will be limitation somewhere. The religion in which our own civilization was cradled is peculiarly unambiguous about it; there is no chance of saying you were not warned; it is there, in the prospectus. "If thy eye, thy hand, thy foot offend thee"—if there is any self-indulgence, any ambition, any love, however honourable in itself, that forms a hindrance to your progress, it has to go. No claim to satisfy a part of your nature must stand in the way of that which is going to satisfy the whole of your nature; there must be no unharmonious tints in the pattern, no discords in the melody. At certain moments in your life, opportunity will present itself, and you will refuse to grasp it, because the suggested course of action does not fit in. Perhaps you will scarcely know why; it will be an instinct, which another man might legitimately set aside as a scruple, but for you it is the voice of duty. States and societies, the mob of men, may be bound by economic necessity, or by their own commitments, to behave as they do; man, the atom, is free to choose. And freedom means, not doing what he likes, but doing what he wants to do.

I must not be understood as prophesying that the contemplation of contemporary affairs will in fact help to implant faith in the new generation, or deepen the soil it grows in. Innumerable factors are concerned; there is a steady policy, all over eastern Europe, of anti-religious mind-conditioning, and we in the West are often frightened of ideas. But, by way of directing men's attention to what religion means and what religion costs, I think the Allied Powers have done the next best thing to not dropping a bomb on Hiroshima. They have dropped it.

SUBLIMATION

X

Brother Atom

THE PATIENT reader who has persevered so far will have become conscious that all I have written could have been written very much better by someone who, in an expressive modern phrase, knew his stuff. My book has, inevitably, pretended to some acquaintance with modern physics, especially where these involve the theory of radio-activity; and the reader will have been in the position of some disconcertingly bright schoolboy who knows that the master is only a lesson ahead of his class. It has indulged in broad generalizations about philosophy; and a turn of phrase here, a want of technical accuracy there, will have betrayed the amateur. But, besides all this, the contents-page is tricked out with words stolen from the jargon of psychiatry; what do I know about psychiatry? And what is the point of these references to traumas, analysis, adjustment, and finally sublimation? Let me confess, once more, that I am trespassing on other people's ground; that I am using cant terms in a popular sense, without having any experience of the new medicine, either as doctor or as patient.

But this I take to be, in broad outline, the theory of the art—that in the mind, as in the body, no wound is more dangerous than the wound which has healed over on the surface and still festers within. The scar may have to be reopened before it can be treated; some experience in the past that lies forgotten—perhaps because we did not want to think about it—must be brought up to the level of consciousness again, dragged out into the light; that is the function of analysis. When this has been done, the patient must somehow be persuaded to see the experience in a truer light, relate it more accurately to the rest of his mental background; that is adjustment. But sometimes, through accident or through design, the whole process may be short-circuited, and the wound healed by a kind of homoeopathic treatment. The urge to which it gave rise in the patient's nature may be canalized in a new direction and to a higher end. That is sublimation.

So far, I have been writing for the benefit (I hope) of people like myself. Of people, that is, whose faith is always more or less at strain in a world where faith is so little set by, and so rarely professed; who are quick, therefore, to scent out the disconcerting element in any new discovery, and cannot be at rest till they have straightened the thing out. Of people who are not sanguine by temperament, yet would fain see justice done, truth apprehended, humanity grow more human and more humane, in this world, without having to wait for the next. Of people whose practical morality is apt to take its tinge, do what they may, from the world around them, and who view with a strange mixture of pious alarm and impious relief any new example which seems to lower the general standard of world-behaviour. I have tried to analyse my own reactions to the scientific view that there is, after all, an indeterminate element at the heart of things; to the prospect of an age in which the possibilities of evil are increased by an increase in the possibilities of destruction; to the news that men fighting for a good cause have taken, at one particular moment of decision, the easier, not the nobler path, I have tried to adjust these instinctive reactions to my general scheme of values, and have offered the reader whatever benefit lies in the working-out of that process.

But it is quite possible that he or she may be of a different temper, and feel that all this preoccupation with the infinitesimal is a morbid tendency; I have been making mountains out of molecules. Indeed, when extracts from this book began to appear in the *Tablet*, a friend wrote to ask whether I had lost my reason, or he his. To some minds, the existence of God is so lucid a conviction that they do not feel the first chilly breath of doubt until the object of faith properly so called—I mean the revelation made by Jesus Christ—is called in question. There are some temperaments so eager for the supernatural adventure that all their hopes are set on heaven; like Noe's dove, they find no rest for the soles of their feet in a world that is transitory. There are some consciences so finely attuned to the music of right living that nothing other people do, or tolerate, has the power to affect them. It is for these more ardent natures that I add this final chapter; indicating as best I can—for I am going beyond my own measure—how to such people the Atom and the Atom Bomb may be, like everything else in creation, what the *Imitation of Christ* calls "a mirror of life, and a manual of holy teaching." They take these things in their stride.

I do not mean that such people will make the mistake—it is surely a mistake—of trying to work out a system of thought which would show revealed Christianity as continuous with the scientific thought of the day, on the lines of Henry Drummond's *Natural Law in the Spiritual World*. That is to hitch your star to a wagon; and though the modern doctrine of indeterminacy may be tempting bait for the defenders of free will, it is to be hoped that the lure will be resisted. A rival error, of taste, this time, rather than of judgement, would be the attempt to put punch into Christianity by re-stating it in "modern" terms. It is not difficult to imagine a popular writer, in the breezy manner, speculating what our Lord's answer would have been if the

God and the Atom

devil had offered him the atomic formula in the wilderness; or refurbishing the Parable of the Leaven by telling us that the kingdom of heaven is like a man bombarding an atom. It is not by providing the opportunity for such theological or rhetorical dexterity that the ideas we have been discussing will appeal to these rare minds. Rather, they will catch up the rhythm and movement of the new thought into their own construction of life, almost unconsciously; the symbolism will match their fearless heraldry, as everything else does.

The truth is, they see the same things as we others, but see them the right way up; see them, that is, from God's end, not from man's. The new discoveries will be, for the inconsiderate, a temptation to pride. That our eyes should have penetrated so far beyond the range of ordinary vision! That the science of numbers, hammered out by the old Egyptians in the course of their star-gazing, should have analysed matter itself! That we should be able to split the unsplittable—does not all this give ground for the boast that we have *done something we were never meant to do*? That man has risen, somehow, beyond human stature, has cheated Destiny? We were meant to go on, surely we were meant to go on, hacking out coal from the seam and pumping up oil and harnessing the rivers to make electricity for us, a little cleverer than rabbit or mole or beaver, but essentially their cousins, with this splendid source of power untapped, unguessed of! And then we set to, and thought; and war came, and made us think quick, and there it was in our hands, the secret we were never meant to find!

All *that* they will see, these people who have vision, the other way up. They will see the gift as one of a series of gifts from God to man. *His* gift evidently, for who, if not he, could implant in nature a power so far beyond all our capacities? Not we, certainly, not blind chance; it is a prize, evidently, awarded by an Infinite Mind to his scholars—his mathematical scholars. His *gift*, it is one of a series, fire, the wheel, gunpowder, the lens, steam, electricity, and so on; and all these hidden away in the great bran-pie of creation, to be fished for by children's hands; every now and then we think we have got to the bottom of it, but always we prove wrong. Should we congratulate ourselves on our cleverness in finding it? Or, on the contrary, strike our foreheads in vexation and call ourselves fools for not having hit upon it earlier? Why, neither; it is part of our destiny; we were meant to find it when we did. We are tapping away, all the time, at the surface of our experience, and there is no certainty when we shall stumble upon hidden treasure. And yet there are dowsers among us, one or two in a generation, people who have a flair for looking in the right place; do we owe our thanks chiefly to these? Or would someone else have made James Watt's discovery, if Watt had not made it?

God and the Atom

Atomic research, to be sure, has been team-work. Yet there was a school-fellow of yours, whose vague memory haunts Eton and Trinity for you, unassuming, untidy, inhibited, golden-good, very likeable; he was killed on Gallipoli, in days when we let manpower go to waste—what if he had died ten years earlier? For he died already world-famous, as the man who established the relation between the X-ray spectrum and the atomic number, Harry Moseley. Would Hiroshima stand, but for him?

God's gift, but a terrible one; all his gifts are terrible. Gold and the beauty of women have made and unmade empires; what blessings wine has called down upon itself, what maledictions! Once God has given us free will, he has given us a wand that turns everything to meat or poison for us. In the intentional order, his gifts can be revoked; the sabbath was made for man, not man for the sabbath. But in the real order, you cannot say whether they were made for us or we for them; they are what we make of them. Here there is no difference between the supernatural order as it is revealed to us, and the natural order as we know it. "Rejection lies in this, that when the light came into the world men preferred darkness to light"; Corner-stone is also, inevitably, Stumbling-block. "The good, the guilty share therein, with sure increase of grace or sin, the ghostly life, the ghostly death"; what is received, is received according to the measure of the recipient. We could not know this without revelation; but we might have inferred it from the conditions of the natural world. When man first made his home into a hearth, he could not guess what havoc was to be wrought, through his malice or carelessness, by fire. When man first rammed powder into a petard, to blast his way through the enemy ranks, he could not guess what useful feats of engineering this sinister ally was to accomplish for him. Impossible to foresee the great fire of London, when man's enemy blew gaps in the huddled house-rows, to stop the ravages of man's friend. On a vaster scale, this new gift, like the old, is ambiguous; it will make or mar the world.

On a vaster scale—we are so much the prisoners of habit that we keep on forgetting what we mean by Omnipotence unless our memories are jogged, now and again, with a fresh revelation of what we mean by power. We travel long miles to see Niagara repeat, with different mathematical determinations, the same miracle of grace and persistency we have noted about the stream in the back garden; mere size impresses us. And now, as if to bring all the fairy-stories true, almost infinite power is to be produced, like the coach-and-six from the pumpkin, out of the almost infinitesimal. The mathematicians themselves have delivered us from the tyranny of measurement, brought us back to David and Golias. May they deliver us from that silly trick of the imagination which still has power to depress us, of conceiving spirit as something smaller than matter. We know that it is nonsense, but we still think of the human soul as something "in" the body, and therefore smaller than the body. Perhaps the very smallness of the atom will make us less subject to the illusion of size.

And at the same time, let us remind ourselves of it once more, this vast strength is concerned, when it is at home, to keep things together; it is destructive only by accident. It has not lain dormant, all these centuries; it has been hard at work, keeping things as they were. The material reflection, you see, and the material coefficient, of that ceaseless act of conservation by which God keeps the universe he has made in being. To keep things together—it is only the illusion of habit that makes us think of destruction as something exciting, preservation as something tame. Obstinately, we mistake news-value for importance.

They do not find, these favoured souls I speak of, that radio-activity, theoretical or applied, comes in to distract them at their prayers. It is only (as it were) a new verse in the hymn of praise which God's creatures sing to him; St. Francis would have fitted in Brother Atom with the rest. It is only a variation, at most, on the old theme; you are not over-excited about a new source of power, when your daily theme is Omnipotence. But what would they tell us about the influence it will have on the world, and the uses the world will put it to? Does their hope, like their faith, remain entirely tranquil? All very well to say that atomic power is a gift we have received from God to make or mar ourselves with; that is true of what theologians call his antecedent will, his absolute will, irrespective of all human reactions. But, since he foresees all the history of our race from eternity, we speak also of his "consequent" will; the design, that is, which he means his gifts to accomplish, in view of the use which, he foresees, we shall make of them. But indeed, there is no necessity for theologizing; the human race is, as a plain matter of fact, about to make or mar itself. And does no shadow of anxiety cloud the prospect of these visionaries; are their eyes too firmly fixed on God to have any tears for human fortunes? If so, we may congratulate them, perhaps, but should we admire them?

If we only knew, we others, what it feels like to have solid confidence in God! Concerned about the future of mankind they certainly are; worried they certainly are not. Certainly they will regard it as possible that we shall have to live through a sort of Wellsian nightmare, which will make the Battle of Britain look as remote, as humane as the siege of Ladysmith looks to us now. If so, it will mean—what? That God has given us chances enough, warnings enough, already, and we are still unrepentant; he is doubling the dose of correction, treating us as he treated Pharaoh when Pharaoh's heart remained stubborn. Perhaps, even, that this is the last act of all; that the terrors which he before us are the direct prelude to those final terrors with which he will come in judgement; we do not know. But, if so, our punishment is at every stage purification, and we must be busy doing penance for our common sins under the impact of a common calamity.... It is perhaps no matter for surprise that some of these people—remember, they feel sin as we do not, feel God's greatness as we do not—jump easily to the conclusion that we stand confronted with the last act of all. But they do not, for that reason, fall into the attitude whose usefulness I questioned in the last chapter but one; an attitude of triumphant despair. A sick world is nearing its death, and, for man or for world, the gateway to death, is suffering, that is all. Leave them to say the prayers for a departing soul, at the bedside of humanity.

All that, for them, is according to schedule; but equally for them it is according to schedule if our present mood of world-disheartenment is only an interlude, only a set-back in a chart whose general curve is upwards. That does not involve any change of attitude. It may be a reign of Christ on earth we have to prepare for; but penance, none the less, is the archway that must greet his coming. In every event, trust in God is the key-attitude these people recommend to us. Let the framework of our lives be as comfortable or as uncomfortable as you will, it will all be according to schedule. That consideration breeds in them the same debonair attitude towards the future which you find, also in the shallow, selfish character who tells you, "It's all a gamble, anyhow." The struggle is more difficult for us, who have enough decency to care, but not enough self-abandonment to trust.

Only they will not be over-enthusiastic when the newspapers predict a time of great prosperity and leisure, while we all sit back and watch the atom do the work for us. That is, perhaps, the greatest danger of all the dangers the atom brings with it; that it may shoot us back, unprepared, into the vortex of prosperity. Let us be honest with ourselves; we are just as far as we were in 1939 from any hope, either of democracy in Europe, or of comity between nations. The world has very largely gone barbarian; and what brings the fact home to us, despite all the polite phrases that are used in public, is the spectacle of that complete material ruin which the barbarism of yesterday brought on Europe, which the barbarism of today hardly aspires, even, to repair. But there is plenty of temptation to say "Peace, peace" when there is no peace, now as in 1938; to condone injustice, to connive at inhumanity, in panic fear of what a fresh appeal to arms would mean; and that temptation would be much stronger, if we could point to a world full of material wealth, and say to ourselves, "Good, civilization has come back to us after all! The dream of the nineteenth century is coming true after all!" If atomic power should manifest the same surprising capacities for making people comfortable, as it certainly has for making people uncomfortable, then we should be in danger of selling our souls to evil, and making up our minds to live contentedly in a world without freedom, without justice, without honour. If we do that, we shall be deserving, and probably earning, a far more grievous chastisement than in the worst days of the early forties.

But I am straying from my purpose, which was only to interpret the attitude of those privileged Christians who will think this book an unnecessary performance, because they will have found nothing in the Atom to give pause to faith, hope or charity. How their faith, how their hope, will take the doctrine of radio-activity in its stride, I have attempted to explain. And I suppose it is still more evident that their charity is not of a kind that can be chilled by any breath of discouragement, just because the discoverers of atomic power have been less than generous in their first use of it, or because the atom itself may be hailed by some minds as the symbol of lawlessness. For theirs are integrated natures, as theirs are integrated minds, and the jar of external shocks only communicates itself to us when there is imperfect equilibrium within. When we find ourselves greedily devouring gossip which helps us to a bad opinion about others, we are *really* thinking that our own record does not show up so badly by contrast; because we are conscious of divided purpose in ourselves, we welcome the evidence of human frailty elsewhere. The integrated nature, knitted together by that strong bond we have spoken of, does not lend itself to such influences; it finds a unity in all things; even what is bad or what is imperfect in the behaviour of others points it, somehow, to the contemplation of the good that has been betrayed, the higher good that has been missed. And in that contemplation it rests undisturbed.

The Moving Finger writes; it is a dangerous business at any time, this offer to interpret the mood of a moment. The journalist writes what will be on the breakfast table tomorrow and in the waste-paper basket the day after; it will not be remembered against him. (I remember hearing of a daily paper which does not keep its files at all after a lapse of two years.) The time-lag involved in the production of a book makes it possible that the mood will have altered almost before your book appears; other things will have caught the attention of the public, new facts, even, may have come to light which alter the whole perspective of the situation. But at the worst, if you have been faithful to your task, you will have put something on record; you will have given the curious historian a document of what the world felt like in September 1945. Meanwhile, the thought uppermost in my mind is one which I have implied rather than expressed in what I have written—that we need an acceleration in the tempo of our spiritual reactions, to meet and match what must surely be an acceleration of tempo in the material developments of history. Difficult not to feel that there are untapped sources of energy in the life of most Christian souls, which need something more than a ready-made formula for their release. I am not advocating world-movements or public meetings; too often, in our time, we have heard well-founded complaints of much cry and little wool. My appeal is rather to the individual conscience than to the public ear; my hope is rather to see the emergence of a Saint, than that of an organization distinguished by initials. For the Saint, whose life supremely realizes the integration I have been trying to describe in this chapter, is, like the atom, incalculable in his moment; holds, like the atom, strange forces hidden under a mask of littleness; affects the world around him, as atomic energy does, not in an arithmetical but in a geometrical ratio—his is a snowball influence. No harm in besieging heaven for the canonization of such and such holy persons now dead. But should we not do well to vary these petitions of ours by asking for more Saints to canonize?

Designed by Fiona Cecile Clarke, the CLUNY MEDIA *logo*
depicts a monk at work in the scriptorium,
with a cat sitting at his feet.

The monk represents our mission to emulate
the invaluable contributions of the monks
of Cluny in preserving the libraries of the West,
our strivings to know and love the truth.

The cat at the monk's feet is Pangur Bán, from the
eponymous Irish poem of the 9th century.
The anonymous poet compares his scholarly
pursuit of truth with the cat's happy hunting of mice.
The depiction of Pangur Bán is an homage to the work
of the monks of Irish monasteries and a sign
of the joy we at Cluny take in our trade.

"Messe ocus Pangur Bán,
cechtar nathar fria saindan:
bíth a menmasam fri seilgg,
mu memna céin im saincheirdd."